A note on variety in documen⸱
nationalities involved in this
conformity in notes, footnotes, ⌄. ⸱
to reflect the diversity of backgrounds from whicʜ ⱳ⸝
scholarly pursuit, however, all vital information is included.

Also from UnCUT/VOICES PRESS

Khady with Marie-Thérèse Cuny. *Blood Stains. A Child of Africa Reclaims Her Human Rights.* **Trans. Tobe Levin. ISBN: 978-3-9813863-0-1**
A ground-breaking memoir by Senegalese Khady, Europe's leading activist against female genital mutilation, forced and early marriage, and unequal gender relations in the African Diaspora.

Hubert Prolongeau. *Undoing FGM. Pierre Foldes, the Surgeon Who Restores the Clitoris.* **Foreword by Bernard Kouchner, founder of Doctors without Borders and former Foreign Minister of France. Trans. and Afterword by Tobe Levin. ISBN: 978-3-9813863-1-8.**
"Can excision be reversed? Can the wounded sex be healed? French surgeon Pierre Foldes ... return[s] to eager patients their sensitivity, femininity, and courage to break the chain. A must read!" – Elfriede Jelinek, 2004 Nobel Laureate in Literature

Preface to the 2012 Special Edition

In an effort to reach a wider audience, we offer you this 2012 special soft cover, limited edition of *Rose Lore: Essays in Cultural History and Semiotics.*

As a student of the Gurdjieff tradition, I have quietly taken note of what Gurdjieff had to say about roses in "Beelzebub's Tales to His Grandson" from the First Series of *All and Everything* (p. 210.) *"In short, both the climate of this country and everything else so delighted the hunters and their families that none of them, as I have already said, had any desire to return to the continent Atlantis, and from that time on they remained there, and soon adapting themselves to everything, multiplied and existed, as is said 'on a bed of roses'."* Some of us know the real meaning of this quote. My translation: we here on earth are so steeped in avarice and warmongering that many of us have forgotten how to love and be easy with each other; at times we seem to be sinking slowly like the ill-fated continent of Atlantis ...

With this new edition, we are reminded again that the rose is no ordinary flower. More and more of us have taken note that it is indeed the most important flower in this universe. The flower has much to teach us about how to live and love peacefully and richly in this life and far beyond this realm. We ought never to stop working for a shift in consciousness. This is the message, the nectar of the major world religions, although fundamentalists and others in self-service have missed and continue to miss the point of living. In this regard, it is also our boon to have a new chapter written by Gamze Demirel, a Turkish scholar affiliated with Suleyman Sah University in Istanbul, who gives us a glimpse of the rose in her rich cultural tradition and classical Ottoman Empire history which includes not only Quranic insights but also shows us how many of the Turkish people have embraced the rose in their everyday life over centuries. Indeed, we thank Fatma Mizikaci, a dear, gentle friend and fine scholar for her assistance in bringing the research of Dr. Demirel to our attention.

Since the introduction of the first edition of **Rose Lore,** the Rose Project has been launched and has begun to speak for itself to carry on the work inspired by this book, quietly, gracefully, up and down the Eastern Seaboard, from Palm Beach to Baltimore to New York. We are grateful to the Rose Team that works quietly to plant seeds to accomplish our rose work. We are especially appreciative of the diverse support of all who have been inspiring and helpful, including but not limited to Toni and Carol Randolph, actor David Keltz, violinist Airi Yoshioko, Meg Ventrudo, Cathy Lowe, Barbara Bell Coleman, Vernon O'Meally, Daryl Dance and significant others, some from far corners of the earth including Fisher Island, Hiroshoma, Istanbul, Frankfurt and New York. We have all grown in our understanding of roses and in deep appreciation of Tobe Levin who sits at the helm of UNCUT VOICES. Indeed the work is never ending.

Frankie Hutton

www.roseproject.com

Table of Contents

Prologue

A cosmic mandala in the shape of a shining white rose was revealed to Dante in a vision. I learned this from Aniela Jaffe, C.G. Jung's secretary and pupil speaking in *Man and His Symbols*. In the many years since reading Jung, I have come across the rose as an icon many times. Why does this special flower serve so well? Perhaps it has a perfect shape that invokes a mandala of wholeness; perhaps we are attracted to its vibrant color and variety of hues; perhaps it is simple beauty that stirs something in our hearts, but whatever the reason, the rose as a symbol has become embedded in our human psyche.

In this collection, remarkable essays explore rose symbology. I propose that you, the reader, attend to whatever special meanings resonate for you, and allow that sensitivity to lead in pursuit of your star. Then, as the mystic Gurdjieff might have said, you will have embarked on a search for meaning that shall be nothing less than just roses, roses.

Seymour B. Ginsburg, Esq.
Boca Raton, Florida

Foreword

Symbols can be extremely powerful. There are animate and inanimate objects, animals, plants, celestial bodies and "things" that have assumed global symbolic significance, sometimes in excess of their intrinsic importance. The exception may be the Sun, the life-giver that is an almost equally powerful divinity for many cultures around the globe, sometimes equivalent in power to its actual status in our solar system. The Moon is another divinity, another all-powerful symbol, appreciated everywhere. Other celestial bodies, three or four of the planets in our system, and a few "fixed" stars such as Sirius, are powerful symbols in their own right.

There are many animal symbols, ranging from fish in general to fishes in particular, to birds of all feathers, to insects and snakes. Some members of the animal kingdom have a symbolic meaning extending far beyond the region they inhabit, for example the lion, naturalized in Great Britain and many other places.

There are inanimate symbols that have metamorphosed into animate, by virtue of their symbolic power, or vice-versa. Such is the case of certain numbers, of certain Arabic numerals, or of metals, the most obvious being Gold and Silver. Religious symbols are all-pervasive: here the symbol denotes a reality of an order far beyond (or above) the importance of the signifier.[1] Obviously, among the religious symbols there is the rose. Among the more abstract symbols we find historic figures and events, which often acquire a parallel history as symbols and then, in turn, influence us, quite out of proportion with their impact as analyzed by the "scientific" "objective" historian. They make up the mythology of modern nations.[2]

In the plant kingdom, perhaps the most important symbol of all is the Tree which, in addition to its almost infinite manifestations, has come to symbolize life. Then there are the flowers which are, botanically rather than symbolically, female sexual organs. If the Sun is symbolically male, or a phallus, it follows that flowers are female, symbolically as well botanically. "Thus with the rose, earlier the flower of Venus, then the flower of the Virgin Mary, from the cup of which many sun heroes, as well as the divine 'rose' child are born."[3] Of course, all flowers function biologically as female organs, but the rose is the most popular floral symbol of the vulva.

The rose is a grateful plant. She will grow almost anywhere, in almost any soil, in almost any climate: she has been observed across the African continent, in the Far East, in the tropical gardens of the Caribbean and South America, and all over Europe. She will grow in the garden, and she will

---

1   *The Encyclopedia of Religion.* Ed. Mircea Eliade. Vol.14. (NY: Macmillan, 1987). 206.
2   See, for instance, Andras Gero, *Imagined History* (NY: Columbia UP, 2006).
3   Marcell Jankovics, *Book of the Sun.* Trans. Mario Fenyo. (NY: Columbia UP, 2001).

grow wild. While her thorns may prick you, -- maybe because she is democratic and a lover of freedom-- she likes to please all passers-by, much as those who once sacrificed themselves on the steps of the temple of Ishtar. She wants to be enjoyed rather than appropriated. She is not demanding. She will open up to the sunlight, to anyone who looks at her, even to the *Little Prince*[4], who searched the globe to find a companion for the lonely rose on his personal asteroid. Voyaging from planet to planet, meeting cranks and people obsessed with money and numbers, it took him a while to realize that down here, on Earth, roses are beautiful but common. Too bad Saint-Exupéry was a racist!

Frankie Hutton has other fine, award-winning works to her credit, but from now on she will be recognized, first and foremost, as the author who has given the Rose more of its due, for it has many dues as an archetypal symbol. This book is a collection, thus the credit has to be shared by those who have contributed to the bouquet—they have recognized and conveyed, along with the poets and the artists, the transcendental, universal values of the symbolism in this super-flower.

Mario D. Fenyo

---

4   Antoine de Saint-Exupéry, *The Little Prince*, Chapter XX.

Introduction

The essays unfolded in this collection are faithful to the indomitable spirit and staying power embodied in the world's most perfect flower, the rose. If semiotics is the study of signs and symbols, then the rose rightfully takes her place at the center of this discipline for it has much to teach us by its example of quiet perfection, beauty, and all around usefulness to mankind. In one of a number of The American Rose Society's books about roses, entitled simply *The Rose,*[5] the point is made cogently that it is counterproductive to try to assign specific varieties to roses in literary or artistic works because such attempts "invariably lead to speculation." The point is also made that the symbolism of a rose can be appreciated "without knowing what specific rose, if any, was intended." The chapters in this collection generally are written with those underlying facts in mind: it doesn't really matter precisely what type of rose comes to mind in order to appreciate the flower as an archetypical symbol in the universe. That said, it should be made clear that it is the rose commonly known as the red tea rose that drives the psyche so to say of this anthology. When studied intently, the rose can be viewed as the universe's way of delivering providential

---

5   The American Rose Society, Beth Smiley, Ray Rogers, eds. *THE ROSE*. (NY: Dorling Kindersley Publishing, Inc., 2000), p. 3.

messages to humans. The messages are not imperceptible, but the real, true messages are well veiled and not for the profane. No doubt that is as it should be.

The chapter-essays contained here then are not about the fragile, cliché-filled, Valentine's Day, commercial aspects of what has come to be the most compelling flower on earth. Of course, it is well known that the rose has found its place symbolically in numerous paintings, family crests, heraldry, coins, flags, dinnerware, clothing, perfume, fabric design and even appears in the realms of architecture and furniture design. Hundreds of thousands of females are named for the flower worldwide. Song and movie titles, paintings, poetry and book titles too numerous to mention have been named with rose in the title. For example, Eva Rosenkranz's book, *The Romance of the Rose*,[6] is a celebration in painting and verse that should not go unnoted as it regards the flower's great variety of uses. From a Jewish metaphysical perspective, another book, Adin Steinssaltz's *The Thirteen Petalled Rose*, is destined to be a classic on the essence of *mitzvoth* or obligations connected with this religious faith. The essays contained in this collection are prefaced with mention of the rose's varied uses and qualities because apparently, as you will witness as these essays unfold, a message is inherent in the flower's flexibility to be used, to serve others. At the outset, we take keen note that the rose has a special mission in the universe; more than one mission in fact. Arcane, cultural and

6   Eva Rosenkranz. *The Romance of the Rose: A Celebration in Painting and Verse*. NY: Prestel Publishers, 2005).

historical aspects are woven into essays presented here for the first time in aggregate. Some inkling of the rose's special mission in the universe can be gleaned from the flower's foremost quality of aptness for human use, crossbred into lovely varieties. The rose actually works hard but quietly for the universe and for humanity, and there is a powerful message in the fact that it does. It provides a cadre of gifts such as wonderful scent, nutrients and pure beauty revealed in its numerous botanical hybrids that are remarkable to touch, to smell and, in some cases, even to taste. In this way it serves humanity and provides the message that each human being must learn to be more flexible, more beautiful and more inclined to serve and to connect with others. The rose is primordial, botanically perfect and is vast in its adaptability as a symbol. Simply put, the flower is user friendly from a number of perspectives, some of which are not well known but will be unfolded in this collection.

Orietta Sala's comprehensive, illustrated guide to 180 varieties of roses from around the world notes that the rose's presence has been recorded 2000 years before Christ and first "appeared on Earth during the Tertiary era, as indicated by the many fossil roses found in various areas of the world— the Baltic region, India, Oregon and Colorado." Sala notes 150 species of roses, 95 indigenous to Asia and 18 to North America. A couple of essays in this volume note that the rose was on earth long before the so called Tertiary era. The American Rose Association provides evidence of fossil records that have tentatively identified roses as 30 to 40 million years old. Intuitively, there is no doubt that the flower predates the arrival of human beings as we know them today. The rose most certainly had a presence in hieroglyphics of Atlantis, if we are guided by illustrations in the cosmological, anthropological pictographs of *Oasphe*, a little known bible of the late nineteenth century that presents insights stretching back for many thousands of years.

Imbedded in mysticism, culture and metaphysics has been knowledge and understanding for eons that the rose is far more than a fragrant, lovely flower with a thorny stem. What the flower implies and lives up to renders it the single most important flower in the universe. The flower has been embraced by secret orders, fraternities, and social clubs. While *The American Rose Society Encyclopedia of Roses* edited by Charles and Brigid Quest-Ritson lists 2000 of the "best rose species and cultivars" available worldwide, the "rose expert" Dr. D. G. Hessayon divides the flowery delight into seven groups: shrub roses, climbers and ramblers, patio roses, floribunda roses, miniature roses, ground cover roses and, of course, the popular hybrid tea rose.[7] A world-wide favorite, the red rose, comes from this latter lot. Perhaps it is not fortuitous that Dr. Hessayon's botanical division of the flower happened to be seven, a number that has great significance in metaphysics.

Although there are now almost endless variations of

7   D. G. Hessayon, *The Rose Expert* (NY: Sterling Publishing Co., 2002), p. 7.

roses as a result of cross breeding, symbolically the red rose is clearly the most revealing of the secrets of the universe. Above, practically out of sight, is the most provocative symbolic evidence of its importance in the rose nebula,[8] known by NASA (the United States' National Aeronautical and Space Administration) as NGC 2244. This vibrant nebula hovers over the earth in the shape of a red rose 4,500 light years away in the cosmos as if underscoring messages of the flower from above for those who've cared to look upward. It brings to mind "as above, so below." This rose nebula cannot be seen with the naked eye, only with the aid of a powerful telescope. Conversely, on earth the botanical rose is at once lovely to look upon and to touch, but has a thorny stem too, sort of like life itself. This commonly called "red rose" has many practical, common uses that are beneficial to humans. It can be used to make rose hip tea and splendid jams and jellies. Rose hips are often partnered with vitamin C for dietary uses. Bulgarian rose oil is even an ingredient in natural progesterone gel.[9] What could be more divine than a flower that is fragrant, edible, hardy, and beautiful while also possessing curative qualities with cultural and metaphysical staying power reaching back beyond recorded history? It seems

---

8   Also known as NGC 2244, the rose nebula can be seen with the aid of a power-ful telescope. It is said to be a stardust sculpture containing globules of dark dust and gas that are slowly being eroded away by nearby massive stars. The rose nebula has been noted in Astronomy 365 Days: The Best of the Astronomy Pictures of the Day Website, edited by Jerry T. Bonnell and Robert J. Nemiroff (NY: Abrams, 2006). Also of interest regarding the rose nebula is Axel Mel-linger's *The New Atlas of the Stars*.

9   www.neutraceutical.com

there's still another message in its centuries of sheer survival. The great 16th century physician, clairvoyant Nostradamus revealed that the rose had vast preventive, medicinal powers that could be used to protect humans from the Black Plague, but this revelation or knowledge came to him too late to save his own family. Nostradamus may not have been the first to be cognizant of the rose's special medicinal qualities for in Native American tribal culture as well as in Hinduism and Indian culture, the nectar of the rose is widely known and has been used for its healing properties and as a symbol of beauty and Spirit for thousands of years.

What's more, worldwide, a number of sacred, obscure documents tell of the rose's talismanic or symbolic meaning. In all its splendor, the rose is becoming better known, but apparently it has always been available as a divine symbol through which those who are awake and aware can understand their true place and mission in the universe. For those who have eyes to see, the red rose is an enduring, providential symbol that bespeaks the work that each aware individual can and must do to connect with Cosmic Consciousness. Hindu mystics knew that symbolism is divine language. They are the "keyholes to doors in the walls of space. It is through these keyholes that we can reach the Qabalistic light and with it draw back bolts that hold domus sancti spiritus into eternity." In some of these essays, we begin to see that this light is possible through living the messages of the rose.

The essays presented here are heartfelt and spew from

diverse perspectives. Here we unveil a collection that stretches across little known history, culture, symbolism and mysticism of the rose. Although many may have taken the flower for granted due to its well-known commercial qualities, a few individuals in all epochs and in a number of geographic locations have known that the rose is a magical and especially beautiful flower beyond its botanical qualities. *The Phoenix*, edited by Manly Hall and thought by many to be the "bible" of early theosophy, reminded readers in its 1931-32 edition that "when a nation ceases to serve the beautiful, it has already begun to die: when a cause departs from the truth, that cause has already failed."[10] The rose is synonymous with beauty in all its manifestations. It is no coincidence that the word rose is the same in English, French, German and other languages and no coincidence that the anagram of rose is Eros, the name for the Greek God of love. This latter fact is a tip off of the rose's significance in the universe, but very few have come to understand the implications of this fact. The rose has been used as an important, quiet symbol in the Qabalistic teachings, in fraternal secret orders such as the Order of the Golden Dawn, in Rosicrucianism, in H. P. Blavastky's *Secret Doctrine*, under veil in the Christian *Holy Bible,* in *The Emerald Tablets*, in *Oasphe* and more. Although its mention is obscure in all of these tomes, it is nevertheless profound as is indicated by an Atlantean pictograph. The Catholic Church has used roses in its rosaries historically, but few Catholics

---

10 Manly Hall, "Introduction" in *The Phoenix, An Illustrated Review of Occultism and Philosophy* (Los Angeles, CA: Hall Publishing, 1931-1932 Edition), p. 5.

know the origins and true history of the rosary and the rose's use in this important iconic symbol.

An important theme of some of the essays is the number of great individuals who have embraced or encountered the rose in their life's work. It's odd and heretofore unrecorded in aggregate that a bevy of unusual, world class men and women were "given" the rose as a symbol in their work or encountered it as an important, sometimes urgent symbol in the course of their lives. In Europe, at various times Nostradamus, William Butler Yeats, Rudolph Borchardt and H. P. Blavatsky knew the rose. In the United States John Ballou Newbrough, C. A. Miles and Edward Bach used it in their extraordinary work. In Japan, Dr. Tomin Hirada was inspired by roses after he treated badly disfigured patients, victims of the Atomic Bomb in 1945. From the gulag of Soviet Russia Daniel Andreev embraced the rose as an important, urgent symbol in his mystical, social philosophy. All of these figures have been quite singular and in most cases performed feats unimaginable for ordinary human beings. How could it be that the rose touched the lives of so many great, divergent men and women from different corners of the globe, from Russia to France, from New Jersey, U.S.A., to Hiroshima, Japan? These people hailed from different walks of life, and the work they undertook was vastly different and unusual, but ironically all had one thing in common through the rose; in some way, they all embraced and aided humanity.

It has become clearer and clearer to all of us who've worked on this collection that the rose is indeed far more than a lovely fragrant flower with thorny stems. Thorns actually symbolize that some messages of the rose must be very carefully considered and guarded and this essential fact will be underscored in a few chapters. Silence, service and sincere work are crucial in order for the rose to appear fully in one's life. Each chapter brings out little known truths and each stands alone in its approach to the rose as an important symbol on earth and even beyond when mystical implications are considered. The rose nebula is the ever present symbolic reminder in the cosmos of what the botanical rose on earth reveals; there is much work to be done and too few human beings are involved in doing it. As a symbol in mysticism, this flower truly requires higher awareness to grasp its full splendor and deeper messages.

Each contributor to this collection brings a unique view and understanding to aspects of the flower. My colleague, Mario Fenyo, professor of history in the University System of Maryland at Bowie State University and translator of Marcell Jankovics's *Book of the Sun*, among other great works, wrote the reflective Forward. Chapter one was written by Michael Price. Michael is Naïve American, Ogibewee, and teaches biology and plant science at one of the tribal colleges in Northern Minnesota; his essay reveals that many American Indians who are part of over 500 tribal affiliations and about 2 million in population knew the power and spirit of the rose long before the White man. Tobe Levin, a humanitarian, scholar, collegiate professor who teaches literature and African-American literature also in the University System of

Maryland in Europe, used the symbol of cutting the rose to bear down on the practice of female genital mutilation. Lisa Cucciniello grew up in the Catholic tradition; she is a teacher, independent researcher and graduate student who traces the origins of the rose in the Catholic rosary. Albert Amao, Peruvian in heritage, now lives in the United States where he is a psychotherapist and teaches metaphysics in the tradition of the ancient Qaballah. Monika Joshi is a registered nurse and former officer in the Indian Army. She was born in Bombay, India, is a licensed Ayurvedic practitioner and now resides in California. She reveals firsthand the East Indian reverence and appreciation for the rose and notes its presence in healing and in Hindu culture. Montgomery Taylor is a former university professor, master astrologer, author and musician who brings a unique rendition to the collection: astrology and the rose. Offering invaluable editorial assistance and translation expertise have been Alexsandr Prodovikov a former Russian economist, Moscow native and Hisae Ogawa, a native of Osaka, Japan, and a world class translator. Both assisted with my Chapter, "Black Plague to Gulag" by providing value translation expertise on great individuals from their native countries.

And now in 2012, with this limited paperback edition, it is my delight to include the research of Gamze Demirel, a Turkish scholar who enlightens us on aspects of the rose deep in Turkish - Islamic tradition and literature. There is so much to learn about the rose and all its esoteric properties. Mystique is a prime quality of the rose that all the contributors have come to know. There is mystique in the way some of us have met and interacted and in some of the rose stories we share that are beyond the scope of this collection. All of us have come to know that the rose is quite singular; there is no doubt it is the universe's quintessential, archetypical flower. We offer this collection gracefully to all who study mysticism, semiotics, herbal medicine and cultural history, for the rose is "planted" in all of these disciplines and beyond.

Frankie Hutton
New Jersey, USA
Original Edition 2007/2012
www.roseproject.com

1

Wild Roses and Native Americans

By Michael Wassegijig Price

I begin this chapter with a story of the Native American oral tradition that tells of the significance of wild roses to the Anishinaabe, the Indigenous peoples of the Great Lakes:

Long ago, the Anishinaabe people were starving after an extended summer drought. The berries were few and scarce, the caribou had moved farther north, and the fish had vanished from the lakes and rivers. Soon, sickness and starvation were upon the village.

One day, a group of young warriors came across Makwa – Black Bear – walking through the forest. The eldest of the young warriors decided that they should kill the bear and bring him back to the village for a feast. In doing so, the young men would be revered as great hunters during this time of need.

As the young warriors crept up on Makwa, the eldest noticed that he seemed to be looking for something. Quietly, they began watching its behavior. Makwa began pulling over long slender thorny bushes and eating the red fruit that was fixed atop. The young warriors watched patiently, waiting for a good shot, as Makwa gorged on the red berries.

Next, Makwa proceeded toward the lake. Feeling a little tense, the youngest of the warriors felt that they should kill Makwa before he got away. As they began to affix their arrows onto the bear, they noticed that he was pulling up the tall water grass and eating the white fleshy bulb beneath the surface of the water. Makwa ate the bulbs while the young warriors lurked in the bushes.

After he had gotten his fill, Makwa walked off into the forest and disappeared. The young warriors quickly ran over to the bushes and ate both the red berries and the whitish bulbs of the tall water grass. They ran back to the village and told the elders what they had witnessed.

From that moment on, Makwa, the bear, was no longer regarded as a food source, but as the teacher of herbs and medicines. Medicine men and women of the Anishinaabe people revere the black bear for its knowledge and wisdom.[11]

The red berries atop the thorny bushes are rose hips, and the white fleshy bulbs of the tall water grass are cattails. This story reflects the in-depth knowledge that Native Americans possess about the plants and animals around them, and the pathways by which they obtained such knowledge. The Anishinaabe, the Native Americans of the Great Lakes region of the United States, have long known that rose hips are the final feast of black bears before they hibernate for the winter. The bears gorge themselves on the fruits in the late autumn when the temperatures began to fall. Soon afterwards, the bears and their cubs retreat to their dens where they will sleep until late April.[12] To the Anishinaabe, herbal knowledge obtained from the bears was considered sacred knowledge.[13] Through observing the behavior of bears in the forest, Native Americans learned that rose hips can be a food source.

This is the story of the wild rose and its connection to the indigenous peoples of North America. Native Americans

---

11   Michael Wassegijig Price, recitation of an Ojibwe story from the oral tradition. There are several versions of the same story that are shared by many Anishinaabe communities.

12   Harriet V. Kuhnlein, *Traditional Plant Foods of Canadian Indigenous Peoples: Nutrition, Botany and Use* (Netherlands: Gordon and Breach, 1991) 248-9.

13   Frances Densmore, *How Indians Use Wild Plants for Food, Medicine, & Crafts* (Washington: United States Government Printing Office, 1928; rpt. NY: Dover, 1974) 324.

have many legends, medicinal remedies and ceremonies, relating to the many species of wild rose, that are centuries old. This thorny shrub is known as a scant food source, herbal medicine and good luck charm. Wild rose can be found across the prairies of the Dakotas, the highlands of Washington, the Smoky Mountains of North Carolina and Tennessee, and the northern woodlands of Minnesota and Ontario. Wild rose has many colorful and descriptive names within the different indigenous languages. Historically, Native American tribes have found many uses for this perennial herb, which we will discuss in this chapter.

Wild Roses, Trade Beads and Tribal Insignia

Many Native American tribes can be identified by their traditional designs and ornamentation. While many plains tribes, such as the Lakota, Dakota and Nakota, used geometric designs in ornamentation, many northern woodland tribes, such as the Anishinaabe and Iroquois in the northeastern United States, and the Cree in central Canada, use floral designs which identify them as woodlands people. Other tribal nations that utilize floral designs are the Blackfeet and Okanagan of the northern high plains, and the Athabascans of interior Alaska. In the early 18th century, European traders brought glass beads and velvet cloth with them and traded with the tribes in exchange for furs. But even before the introduction of glass beads, many tribes had already developed intricate floral patterns and utilized items such as dyed porcupine quills, bone beads, and colorful cordage to decorate their outfits. The introduction of glass beads quickly transformed the appearance of traditional regalia for many tribal communities who traded with the European explorers. By the late 1700s, Anishinaabe people had already incorporated vibrant colorful beadwork and black velveteen cloth into their traditional attire.[14]

The Anishinaabe use woodland floral designs to ornament their dance outfits, bandolier bags, cradleboards, and various other ceremonial objects. The floral designs reflect the local plant species that they were most familiar with. The most popular designs for Ojibwe beadwork are wild roses, rosebuds, bluebells, maple leaves, acorns, and wild grapes, and are used as much today as they were over two hundred years ago. [15]

The author of this article has sat in on conversations with Anishinaabe elders who stated that originally the floral designs represented medicine plants. Wearing the design of a particular medicinal plant could bring power and good medicine to that individual.[16] These floral designs can be seen today at powwows and ceremonial gatherings all across Anishinaabe Country.

---

14   Carrie Lyford, *Ojibwa Crafts*. (Wisconsin: R. Schneider, 1943, rpt. 1982) 145.
15   Ibid. 147.
16   Michael Wassegijig Price, recollection of casual conversations with women tribal elders from the Leech Lake Band of Ojibwe in northern Minnesota. The passing of knowledge from one person to another is known as the "oral tradition" and is a custom that is centuries old.

Wild Roses in Native American Languages

Native American languages are as diverse as the landscape of Turtle Island – a metaphor for the North American continent meaning that the land is actually on the back of a giant turtle. Many indigenous tongues share a root with many dialects. The major language groups across the North American continent are Siouan, Algonquin, Tewa, Diné and Iroquoian.

In general, Native American languages tend to rely heavily upon metaphoric and symbolic relationships. Many descriptions and abstract concepts are woven together using elements of the local environment and cultural traditions relative to each tribal community. One study attempted to categorize the indigenous naming of wild plants, but found that the indigenous ways of knowing plants were far different from the scientific Linnaeus method of categorization.[17]

The Anishinaabe People comprise many Native American tribes including Ojibwe, Ottawa, Pottawatomie, Menominee, Meskwaki, and Shawnee. The language of the Anishinaabe is placed by linguists in the Algonquin class, a group that includes many tribal peoples from Minnesota to Maine, from Indiana to Hudson's Bay, Ontario.

In the Ojibwe language, spoken by the Anishinaabe people, the name for wild rose is *Oginiiminagaawanzh*

---

17   Iain J. Davidson-Hunt et al. "Iskatewizaagegan (Shoal Lake) Plant Knowledge: An Anishinaabe (Ojibway) Ethnobotany of Northwestern Ontario." *Journal of Ethnobiology* 25(2) (Fall/Winter 2005) 189.

(pronounced O-GINEE-MINA-GAW-WUNZH). An analysis of the word reveals:

ogini – his mother; *imin* – fruit; *agaawanzh* – small bush

The translation is *"Mother Fruit from a Small Bush."* The name refers to edible fruits, or rose hips, that are found at the tips of the rose stalks. This name also refers to several species including Virginia Rose (*Rosa virginiana*), Prickly Wild Rose (*Rosa acicularis*), Smooth Rose (*Rosa blanda*) and Swamp Rose (*Rosa palustris*). All of these species are indigenous to the northern boreal forests and peat lands of the Great Lakes region.[18], [19]

The Anishinaabe word, *Ogin*, has a more contemporary translation which means *"tomato."* Tomato and rose hips are both edible fruits from each respective shrub. The multiple meanings of this word make translation complex and challenge understanding the naming of this flowering plant.[20] , [21]

The Meskwaki people are a small tribe of Anishinaabe that reside in central Iowa. Meskwaki translates to *"People of the Red Earth,"* possibly referring to the deep red, iron-rich clay soils of the region. These southern relatives of the Anishinaabe call the wild rose *"Kishipi'iminaki"* which means

---

18 James E. Meeker et al. *Plants Used by the Great Lakes Ojibwa* (Wisconsin: Great Lakes Indian Fish & Wildlife Commission, 1993) 82, 225, 394.
19 John Eastman, *The Book of Swamp and Bog: Trees, Shrubs, and Wildflowers of Eastern Freshwater Wetlands* (Pennsylvania: Stackpole Books, 1995) 155-8.
20 Fredric Baraga, *A Dictionary of the Ojibway Language* (St. Paul: Borealis/Minnesota Historical Society Press, 1878) 317.
21 John D. Nichols and Earl Nyholm, *A Concise Dictionary of Minnesota Ojibwe* (Minneapolis: U. of Minnesota P., 1995) 238.

"to itch like hemorrhoids" referring to the bowel irritation caused by ingesting the microscopic hairs found on the seeds within the rose hip fruit.[22]

The Anishinaabe have a different name for the Wild Prairie Rose (*Rosa arkansana*) which is named *Bizhikiwiginiig* (pronounced BIZHIKI-WIG-INEEG). An analysis of the word reveals:

bizhiki – buffalo; *gin* or *ogin* – rose flower; *iig* – plural ending

The translation means "Buffalo Roses." Buffalo once roamed the tall grass prairies of the high plains. Their hooves tilled the soil and created landscapes suitable for prairie grassland species. The buffalo and grasslands evolved together over the centuries. But rarely did the buffalo venture into the eastern woodlands of Minnesota and Manitoba. The Anishinaabe associated the Wild Prairie Rose (*Rosa arkansana*) with the buffalo's natural habitat.[23] James Garrett, Ph.D., Lakota tribal member, and Professor of Bison Rangeland Science at Little Hoop Tribal Community College, says that the buffalo graze on the wild rose leaves, stalks and flowers throughout the summer and autumn, and rose hips throughout the winter months.[24]

---

22   Huron Smith, *Ethnobotany of the Meskwaki Indians* (Milwaukee: Bulletin of the Public Museum of the City of Milwaukee, 1928) 242.
23   Meeker ibid. 53.
24   James Garrett, Interview by the author. Rapid City, South Dakota. March 25, 2007.

The tribes of the Great Plains speak a different language from the Anishinaabe People of the woodlands. The Plains tribes have a strong connection to the tall grass prairies and the buffalo. Tribes of the Great Plains such as the Dakota, Omaha, Pawnee, Cheyenne and Blackfeet all have names for the various rose species in their respective languages:[25], [26]

Tribe	Tribal Name	Latin Classification
Dakota	Onzhinzhintka	*Rosa pratincola*
Ponca-Omaha	Wazhide	*Rosa pratincola*
Pawnee	Pahatu	*Rosa pratincola*
Blackfeet	Kinii	*Rosa acicularis*

The presence of wild roses on the plains inspired the naming of the Rosebud Sioux Reservation in South Dakota. Through the Indian Reorganization Act of 1934, Rosebud Agency was established in Kyle, South Dakota, to administer governmental services to the Sicangu Band of Lakota. The name "Sicangu" (pronounced SEE-CHONG-GOO) is a Lakota word which means "burnt thighs." The Rosebud Sioux reservation got its name from the charismatic Lakota leader Spotted Tail and the not-so charismatic governmental officials who visited the region in 1877 and chose it as the site of the new Lakota territory. Spotted Tail and other Lakota noticed the prevalence of wild roses growing across

---

25 Melvin R. Gilmore, *Uses of Plants by the Indians of the Missouri River Region* (Washington: U.S. Government Printing Office, 1919; rpt. Lincoln: U. of Nebraska P., 1977) 33-4.

26 Roslyn LaPier, Interview by the author. March 15, 2007, Browning, Montana.

the prairie landscape. The wild roses were reminiscent of the Rosebud River Country in present-day Montana, where they traditionally camped while en route to the Little Big Horn Country. The name "Rosebud" memorialized a part of their history to which they could never return.[27]

Wild Roses and Native American Spirituality

Native American spirituality is a complex weaving of the physical world with the spiritual world. It recognizes that all living things possess a spirit, including animals, plants, rocks and rivers. By recognizing that these things have a spirit, Native peoples developed ways to commune with the spirit world that surrounds them. These relationships have evolved throughout centuries of harvesting foods, conducting ceremonies, practicing traditions, and storytelling.

The spiritual landscape of Native Americans is as diverse as the physical landscape that they inhabit. Ceremonies, herbal knowledge and cultural practices are directly related to the Eco region and homeland of a particular tribe. The ceremonies practiced by Pueblo peoples of the desert southwest would have little meaning to the woodland people of the northern woodlands, and vice versa. Buffalo ceremonies conducted by the plains tribes were practically unknown to the Cherokees of the Smoky Mountains. Though many tribal nations share

27   James Rattlingleaf, Interview by the author. April 23, 2007, Kyle, South Dakota.

certain customs and beliefs, describing the spirituality of Native Americans as one group is difficult, if not impossible. Many traditional tribal peoples believe that wild rose has a unique spirit that is deserving of respect and recognition. Before any part of the plant is harvested, permission is asked from the spirit of that particular plant, and a gift is usually given in gratitude for its harvest. Tobacco is usually a gift given to appease the spirits. By performing this ceremony, Native peoples humble themselves and take only what is needed for their use. This worldview has sustained indigenous communities with their environments for centuries.

The meaning and purpose of many ceremonies were never shared with Western society. When the U.S. Government in conjunction with Christian churches attempted to force assimilate young Native children into English-speaking Christian society in the early 20th century, many ceremonies and tribal traditions went underground and were held in secrecy for generations. Some ethnographers and anthropologists recorded tribal languages and customs, but one key point to remember is that their writings were interpreted through the eyes of the ethnographer, not by the indigenous peoples who created them. For instance, the Omaha practiced a ceremony by rubbing crushed rose petals on their hair. There were many assumptions and theories as to the importance of this practice, and many assumed that it was used as a hair perfume. The Omaha never shared the purpose of this ceremony with the outside world and the rite continues to be held in secrecy.[28]

For many Native American tribes, smoking was a ceremonial activity that created kinship and goodwill between people, communities and nations. Pipe ceremonies were conducted between peoples seeking relationship or common alliances. Ceremonial pipes were carved from the sacred red pipestone (*Catlinite*) found only in the area of Pipestone, Minnesota. Tribal peoples made smoking mixtures from different plants indigenous to their homelands. The Plains tribes such as the Omaha, Pawnee, and Dakota smoked the inner stem bark of wild rose in their pipes, or they mixed it with the inner bark of red willow and tobacco.[29] The Klamath people of Oregon used wild rose stems as pipe stems because they became rigid when dried and could be easily hollowed out.[30]

The Okanagan-Colville people proudly called themselves "The People of the Plateau" because their ancestral homelands are located in the eastern highlands of Washington State. Okanagan spiritual leaders made good luck charms by boiling wild rose stems in water and then sprinkling the brew over fishing lines and nets for a successful catch. Hunters sprinkled the water charm on themselves for good luck which also served to diminish human scent for hunting. This water charm was also used to chase away bad

28   Melvin Gilmore, *A Study in the Ethnobotany of the Omaha Indians* (Nebraska State Historical Society Collections, 1913) 314-357.
29   Gilmore, *Uses of Plants.* 33-4.
30   Frederick V. Coville, *Notes on the Plants Used by the Klamath Indians of Oregon* (Contributions from the U. S. National Herbarium, 1897) 87-110.

spirits or prevent bad medicine inflicted by another person. Okanagan medicine men would also use the branches of wild rose to sweep out a freshly dug grave to prepare it for burial ceremonies.[31]

Wild Rose as Food Source

Every part of the wild rose plant, which includes stems, roots, leaves, hips, buds, was used as either a food source or food additive. Rose hips were harvested in the autumn or throughout winter. Hips were gathered after the first frost because freezing temperatures made the fleshy tissues tender, though the ratio of fleshy fruit to seeds is small. Though not a staple food, rose hips were considered "famine food" because they could be found above the snow during the winter months when starvation was most prevalent. The leaves and flower petals are eaten raw throughout the growing season, while the roots, stems, leaves, petals and hips were all used in making teas.[32]

Many tribes knew the food value of rose hips, but they were also aware that the fine microscopic hairs on the seeds could create bowel irritation. A coyote story shared by many tribal communities warns of eating too many rose hips:

> *Once Coyote ate the bright red fruits of the wild rose. The next day, his anus began*

31 Turner et al. *Ethnobotany.* 131.
32 Kuhnlein, Ibid. 248-9.

to itch. It itched so much that Coyote was desperate to seek relief. Coyote grabbed a handful of willow bushes and began wiping himself. He wiped himself so much that he began to bleed, and he bled all over the willow branches. Still, he could not stop the itching. That is why the branches of the Red Willow (Salix laevigata) are bright red.[33]

Anishinaabe elder and scholar, Basil Johnston, tells a story of how Nanaboozhoo, the Trickster god of the woodlands peoples, gave thorns to the wild rose to protect itself from the gluttonous rabbits. For nearly decimating the wild roses from the landscape, the animals ganged up on rabbit, grabbed him by the ears which stretched them out long, and punched him in the mouth which split his upper lip.[34] This story demonstrates the importance of the wild rose as a food source for other animal species, and the physical characteristics of the rabbit are reminiscent of this legend.

Other relative species of the wild rose (Family: Rosaceae) which are traditional food sources for Native Americans are Blackberries, Raspberries and Smooth Juneberry. The Ojibwe names are as follows:[35]

---

33   Michael Wassegijig Price, recitation of a traditional story shared by many Native American communities. There are several versions that exist for this particular theme.

34   Basil Johnston, *Ojibway Heritage* (Toronto: McClelland and Stewart, 1976; rpt. 1990). 44-5.

35   Meeker, Ibid. 29, 125, 231.

Common Name	Ojibwe Name	Latin Classification
Blackberry	Odatagaagomin	*Rubus allegheniensis*
Raspberry	Miskomin	*Rubus idaeu*
Smooth Juneberry	Gozigwaakomin	*Amelanchier laevis*

Wild Rose as a Medicine

Wild Rose contains significant amounts of vitamins C and A, minerals such as calcium, phosphorus and iron, and it can act as an antioxidant. "*Wild Rose is the super-plant of the Blackfeet people,*" states Roslyn LaPier, Blackfeet tribal member, Tribal Ethnobotanist and faculty member at the Piegan Institute in Browning, Montana. "*Every part of the plant is used: roots, leaves, flowers, hips and buds.*"[36] Rose species can also act as an antiseptic, astringent and diuretic. Throughout the centuries, tribal communities shared its medicinal remedies, while other groups used the plant in entirely different ways.

Many Native Americans, including woodlands and plains tribes, used poultices of various wild rose species for eye mucous irritation and cataract inflammation. People of the northern tribes such as the Ojibwe, Cree, and Iroquois occasionally suffered from snow blindness and eye soreness during the winter months. Healers from these communities created an eye wash from the petals, stem bark and root bark

---

36   Rosyln LaPier, Interview by author, March 2007, Browning, Montana.

of wild rose to relieve eye pain and irritation.[37] The Ojibwe, in particular, treated cataracts by creating two eye washes, one made from the root bark of wild rose and the other from the root bark of wild raspberries, and applied them separately and successively. The wild rose eye wash treated the irritation, and the wild raspberry eye-wash worked to heal the tissues around the eye. This eye-wash treatment was administered three times a day until the eyes healed.[38]

Teas were also used to treat diarrhea and gastrointestinal problems. The Meskwaki people of northern Iowa created a decoction of the fruit to treat hemorrhoid inflammation, although the microscopic hairs from the rose hip seeds had to be carefully removed because, as mentioned earlier, it too would cause bowel irritation.[39] The Blackfeet of northwestern Montana made necklaces with dried rose hips and used a concoction of wild rose stems and root bark to treat children and adults with diarrhea.[40] The Anishinaabe people used dried rose petals as a medicine to treat heartburn.[41]

In the Plateau region of Washington State and British Columbia, Prickly Rose (*Rosa acicularis*) and Dwarf Rose (*Rosa gymnocarpa*) grow abundantly. Okanagan people call wild rose "Coyote Berry" because they observed the coyotes eating rose hips. For bee stings, Okanagan healers

---

37   Kelly Kindscher, *Medicinal Wild Plants of the Prairie* (Lawrence, Kansas: University Press of Kansas, 1992) 189-93.
38   Densmore, Ibid. 292.
39   John Eastman, *The Book of Swamp and Bog: Trees, Shrubs, and Wildflowers of Eastern Freshwater Wetlands* (Pennsylvania: Stackpole Books, 1995) 155-8.
40   Kindscher, Ibid.189-93.
41   Smith, Ibid. 385.

placed chewed leaves on the sting area to reduce swelling and irritation of the skin.[42] People of the Shuswap First Nations (Thompson Indians) in British Columbia treated athlete's foot by placing rose leaves in their moccasins.[43]

The ancestral homelands of the Cherokee people are in the Smoky Mountains of western North Carolina and Tennessee. Here, the Virginia Rose (*Rosa virginiana*) is the predominant wild rose species. The name for wild rose in the Cherokee language is "*Jisdu Unigisdi*" which translates to "*what the rabbits eat.*" Cherokee healers created a "medicine bath" using the wild rose roots to treat babies and children with worms.[44]

Many Native American tribes used wild rose in the treatment of open or bleeding flesh wounds. The Ojibwe created a decoction of pulverized wild rose roots to prevent bleeding by applying it directly to the wound. The Crow of central Montana created a vapor using crushed roots in boiling water to treat nosebleeds. They also used the same remedy to make a compress to reduce swelling of skin wounds.[45] For treating burns, the Pawnee harvested the large hypertrophied outgrowths on the rose stem called galls. Galls are tumors that form on the stem as the result of a bacterial infestation. Healers would char the wild rose galls in a fire, crush it into

---

42   Turner, Ibid. 131.
43   Ibid. 267.
44   Linda Averill Turner, *Plants Used as Curatives by Certain Southeastern Tribes* (Cambridge: Botanical Museum of Harvard University, 1940). 29.
45   Jeff Hart, *Montana Native Plants and Early Peoples* (Helena: Montana Historical Society Press, 1976; rpt. Montana Historical Society and Montana Bicentennial Administration, 1992). 35-6.

a poultice, and apply it directly to the wound.[46] In a similar fashion, the Paiutes of the Warm Springs Reservation in Oregon chewed the wild rose galls into a poultice and applied it to boils.[47]

The spirit and essence of wild roses have permeated Native American cultures for centuries. This small thorny shrub has inspired healers to cure the ill, quelled hunger during times of struggle, created funny stories and fables that would live for generations, nourished the body, and lifted the spirits of the oppressed. Like the wild rose, Native Americans have persevered in the harshest of environments, adapted to rapidly changing surroundings, survived adverse conditions, and yet have maintained their beauty and essence. The wild rose possesses much power, magnificence and elegance. Human beings have to only open their minds and souls to discover its medicine.

This article is dedicated to my mother, Rita Wassegijig (Bright Sky), who passed on to the spirit world on May 5, 2002. Since childhood, her favorite flower had been the wild rose.

46   Gilmore, Ibid. 33-4.
47   James Michael Mahar, "Ethnobotany of the Oregon Paiutes of the Warm Springs Indian Reservation" (M.A. Thesis, Reed College, 1953) 25.

2

FGM – Or Cutting the Rose in Alice Walker's Garden

By Tobe Levin

Many have written of genital mutilation, and many have denounced it [including many respectful of tradition]. ... I feel enfeebled here, unable to add a constructive ... insight, depressed by the persistence of a repulsive "rite," and made small, as we all are, by capitulation through inertia. Genital cutting is an extreme abuse of human rights. Like slavery and apartheid, it is unacceptable. How can we stop it? By talking about it with angry, unbitten tongues. By never forgetting about it, and by not letting the issue slide back into obscurity now that we have learned of its pervasiveness and tenacity.

Natalie Angier. *Woman. An Intimate Geography.* (88)

In "The Cut," Maryam Sheikh Abdi courageously tells us what it was like the day she faced the knife. As the children were taken one by one, the six-year-old, nauseous, waited with "ears blocked" by a single sound, the shrieks and sobs of "wailing ... girls."

Then, it was her turn.

Obediently, I sat between the legs of the woman who would hold my upper abdomen, and each of the other four ... grasped my legs and hands. I stretched apart [with] each limb firmly held [until], under the shade of a tree ... the cutter began ... [ellipses in original]

"To this day," Maryam proclaims, decades later, the wound remains fresh: she had screamed until her "voice grew hoarse, [and no] cries could come ... as the excruciating pain ate ... flesh."

As in *Born in the Big Rains. A Memoir of Somalia and Survival*, where Fadumo Korn, in a near death experience, floats above a scene of carnage, Maryam faints, only to awaken to "unbearable," "agon[y]" as well as "a slap across [the] face" for cowardice. She had never ceased writhing. And even though the surgery itself – the cutting and stitching -- was over, "the pain kept coming in waves, each wave more pronounced than the one before it ... for the blood oozed and flowed," attracting the "scavenger birds [who] move[d] in circles and perch[ed] on nearby trees."

What happens next? Remarkable for its programmatic character, the narrative poem spells out in unwonted detail what the 'initiate' endures. After "the bloody sand" is "scrubbed" off, a hole is dug, and the *malmal* is pounded. It will be "pasted where [the] severed vaginal lips had been" before the legs are tied. Over a fire containing "dried donkey waste" and herbs, Maryam sits, only to hear "the blood dr[i] pping on the charcoal." Will she bleed to death?

She survives, bound in camel hide, and is instructed how to sit, stand, and shuffle so as not to disturb the "sealing" of "that place," now a source of endless torment. Excretion is hell. As she lies on her side, the urine, "more burning than ... the razor," leaks out drop by drop. Unwashed and undried, the wound flames for hours.

The healing takes a month during which time the diet is strict: no vegetables, oil, or meat, and very little water. Nor do the victims bathe. Hence, the lice attack. Nesting "between the ropes and our skin," the insects bite; the wounds itch. Respite? There is none.

Nura Abdi, together with her age-mates, goes through it, too. After losing consciousness, she recovers to see

> blood on the floor and ... what had been sawed off all of us.... tossed ... in a pile. .... Later I learned that someone had dug a hole and buried [those parts] somewhere in the courtyard. Exactly where we were never to learn. "What do you need to know for?" was all they would say. "It's long gone to where it belongs. Under the earth." (33)

With exceptional courage, Maryam, Fadumo and Nura testify to their infibulations, a genital assault that culminates in burial of the "offensive" body parts. Out of sight, out of mind.

Cutting the Rose

Although the late Monique Wittig, in a post-modern quip, claimed to have no vagina, thereby removing herself from the class of candidates for genital clipping, individuals whom patriarchal binaries identify as clitoris-carriers too often undergo ritual sexual attack. The figures are well-known and increasingly horrible because failing to decrease significantly.[48] Despite campaigns that began in the 1970s, official estimates of victims still cite more than 130,000,000 who have undergone or are threatened with the treatment described above (15% at risk of infibulation, the remaining 85% of clitoridectomy or excision).[49]

---

48 Parrot and Cummings offer "Success Stories and Promising Practices" recording victim reduction in certain ethnic groups in Senegal, Uganda, Kenya and Burkina Faso as a result of targeted interventions or best practices. But the numbers saved remain (too) small.

49 See "Female Genital Cutting and the Demographic and Health Surveys." *FGC Data from DHS Surveys, 1990-2004.*
http://www.measuredhs.com/topics/gender/FGC-CD/start.cfm#DHS%20data

Several questions obviously follow. Why are female genitals so treated, without reciprocity?[50] Why is the custom so tenacious? And is this 'merely' an African problem?

More than 200 years ago, English Romantic poet William Blake captured centuries of hostility to women's sexuality in "The Sick Rose":

O Rose, thou art sick!
The Invisible worm,
That flies in the night
In the howling storm,

Has found out thy bed
Of Crimson joy;
And his dark secret love
Does thy life destroy.

A metaphor for female genitalia, the fragile flower signals pain and decay throughout Western literary history. Superficial and transient in its splendor, the rose attracts the worm, a vile and sepulchral penile metonymy, implying lethal male desire linked to fear of sex. While consummation in the "howling storm" endangers both participants, the male's *angst* translates into violence, and he neuters the temptress.

---

50   The least damaging intervention is clitoridectomy and yet Angier's words apply: "The unarguably vile practice goes by various names, including female genital mutilation, or FGM; African genital cutting; and female circumcision – although as many have pointed out, it is more akin to penile amputation than to male circumcision and should not be given the courtesy of comparison" (86).

Faced with a treacherous organ, seductive yet smelly, filthy, and disgusting, men who refuse to marry open – loose – girls incite to genital modification. 'Logic' dictates that the offensive parts must go.

Aware of this cultural encoding, contemporary artists have revived the rose. In Stuttgart, for instance, on International Women's Day 2006, performance artist Dorothea Walter staged the underlying tension between wounding and recovery in a sketch titled "Love the Rose. On FGM" (Gruber 13). The theme also inspired creative staff in European advertising agencies, answering a call for visuals by MEP (Member of the European Parliament) Alexander Alvaro. In *DON'T. Women's Art Protest against Female Genital Mutilation*, Katja Kamm adorns with contented female figures the uncut undulation of an upright bloom. The caption: "Let us worship our female nature ... no more FGM" (11). "It's my nature. Don't destroy it" is likewise Elke Ehninger's plea (7).

Not limited to Western portrayals, however, the two opposing themes, affection for the "(American) beauty" and prophylactic floral trimming, are equally captured in works of art protesting FGM[51] by Nigerian artist Godfrey Williams-Okorodus. In "Defiance of Pain 1" and "Defiance 2" a golden rose dominates the canvas, yet the paintings differ as to the victims' options. In the first, sharing the visual center with a naked razor, the flower is grasped by a small girl, her back turned to a wall of women. Inexorable, they stand for custom, as not a chink appears in the barrier they form. Keeping her blossom is out of the question, despite the suggestion of "defiance" in the painting's title. Strikingly pale against the adults' deep hue, seated on the ground while grown-ups stay rooted in tradition, the young one is outnumbered and without allies. Can she rebel? Or is the artist alone insubordinate in visualizing the suppressed (Reinharz et al 12)?

"Defiance 2" gives the girl a better chance, for here she dominates the foreground, arms crossed in rebelliousness. A flower next to her also glares in rich, red fullness, not yet faded like its companion bloom, while the village recedes behind both audacious blossom and evading teen. Will she escape? Possibly. But wait! There, to the left, is a palimpsest, a ghostly rendering of two imams, and a third hovering just

---

51 "Through the Eyes of Nigerian Artists. Confronting Female Genital Mutilation" is an exhibition of paintings and sculpture that opened in 1998 in Nigeria, and was then shown at the Women in Africa and the African Diaspora (WAAD) conference at the University of Indiana-Purdue University. The artworks came to Germany in 2000, where 70 venues displayed them over the next six years. In the USA, the exhibition has been viewed at Brandeis, Harvard, Cornell, Bucknell, SUNY-Fredonia, Bridgewater State College and Monmouth University. It is available for rent from UnCUT/VOICES Press. You can see a slideshow of the Brandeis opening at
http://share.shutterfly.com/action/welcome?sid=0BbNmTVuzYsXUQ

off the margin, leading the story off of the canvas and back into the world. The men appear at once behind and in front of the girl. Blocked again (Reinharz et al 13).

Roses suffuse a third Williams-Okorodus oil, "The Uhrobo Bride" (Reinharz et al 12). In the "Defiance" series, women are clearly associated with the genital attack, but the reason for female complicity goes unaddressed. The aptly titled "Bride" associates the amputations with enhancement of beauty and desire. Four elements stand out in the bust of a serene woman with harmonious sensual features: the same passionate crimson triangulates full lips, beaded necklace, and fat flower. The single discordant element, a razor adorns her right ear – the instrument aestheticized as jewelry. The same elision of disfigurement and fashion also frames the portrait: ten palms in an ambiguous gesture (stop? I give up?) beckon to the viewer, most tattooed with flora but one with a blade. Here Williams-Okorodus anticipates the "truth" most challenging to activists: women's emotional attachment to the damage, justified as pain in the service of beauty. As the French say, "Il faut souffrir pour être belle."

Is FGM then 'mere' cosmetic surgery? It is ethically unacceptable, of course, because forced on minors, but, analyzing "Female Circumcision among Nubians of Egypt," Fadwa El Guindi asks: "'Had *This* Been Your Face, Would You Leave It as Is?'" (Abusharaf). Is it an aesthetic issue?

In large measure, yes, and for that reason resistant to arguments from health or rights. Motivated by idealism and their wounds, Maryam, Nura and Fadumo make valiant efforts to overcome the 'modesty' their upbringing taught them to value. They are indeed torn; for their emotions tell them that their suffering was not for naught, that the ordeal had done them good. Nura is shocked to find herself in Germany among a nation of the unclean. And Korn admits her loathing for genitals left in the raw.

'You stink', Fadumo, age ten, tells two Spanish sisters while their parents sip cocktails at her uncle's Mogadishu home. In this scene, Somali girls confront their European classmates, taunting them for the uncouth status of their nether parts, open, wet, or worse. 'You're dirty, yuck!' Fadumo sneers. 'You drool! You're going to hell!' (119) As children will, the Spanish youngsters lie: 'We're circumcised like you!' But Fadumo declares, 'I don't believe it' and escalates doubt into assault. Once the girls have retreated to her bedroom, Fadumo challenges her guests. 'Prove it! Show us!' (120) When the Spanish girls don't, the Somalis do:

> Determined, I undid the button and unzipped my jeans, hesitated for a fraction of a second, then down went the pants over my bottom and I pulled aside -- not for long but long enough for everyone to see -- my panty crotch. One after the other, everyone did the same. (120)

Everyone, that is, except the Spanish girls whose reluctance provokes attack:

'Get ,em. Let's look for ourselves!' As if waiting for my command, the gang stormed, threw the sisters on the bed, ... and pinned their arms and legs while we removed their underpants. 'I knew it,' I cried. 'Look at them! How wrinkled they are, how shriveled, how ugly. Yuck!' (121)

First, we find assailants who have themselves been seized, had their legs torn apart, and been constrained by a regiment of women. This, then, is learned behavior. Second, the anecdote conveys a point that no discussion of the topic should neglect. Not only is FGM *normal* to those whose ethnicity mandates it as a condition of gender identity and group membership, but beauty and cleanliness[52] are thought to result. These in turn give the girls pleasure – if they recover, that is, -- as they have longed for the event, to change status and become "women." Genital erasure is the sign of belonging -- as only cut roses can adorn the home.

Hence, the scene reveals the practice's psychological appeal and its motor in peer pressure. To become a "positive deviant" then means making a sharp turn-around from pride to shame to an emotionally neutral knowledge from which activism may emerge. But this is a burdened mental move and takes *enormous* valor.

---

52   Cleanliness not only as hygiene but also as (ritual) purity. See Mary Douglas on "how rituals create and control experience" (82).

The needed audacity is not easily found. At Mt. Holyoke in 2004, encouraged by the screening of Ousmane Sembene's ground-breaking film *Moolaadé*, I spoke on "Somali Immigrant Memoirs and Campaigns to End FGM in Germany." After the talk, two U.S.-born Somali students approached. The speech, they said, was "awesome." Would I give it again for the African and Caribbean Students Association? Of course, but on condition that we work together. So the three of us met several times to study and prepare presentations for one or two of the local prep schools. Thanksgiving, however, brought a change of heart. "We're really sorry," the girls said, "but our mothers have forbidden us to work on this." Why? If word should get back to the immigrant community, they would have been denounced as traitors.

A bouquet for Alice Walker

Sensibilities can indeed be tender on this subject, as Alice Walker found out in the USA.

In the early 1990s, Walker became the first personality of world renown to publicly oppose female genital mutilation. Her novel *Possessing the Secret of Joy* (1992) and documentary (with Pratibha Parmar) *Warrior Marks* (1993) did more than decades of international and grassroots agitation by ‚ordinary' citizens to raise the issue with law- and policymakers in Africa and the African Diaspora. Before the 1990s, FGM could still be considered a taboo subject, the number of academic and popular articles remaining shamefully small. True, in 1982 Elizabeth Passmore Sanderson published an 82-page bibliography, but ethnographic work, the bulk of listings, addressed academics, not legislators, citizens, activists or practicing groups. And anthropologists, to put the best construction on it, although admirably aware of colonial abuses, tend to be blinded by good-will – toward systems in power, which means patriarchal privilege over women.

Walker, however, is not an anthropologist but a creative writer whose intervention was followed by action. I contend it is more than coincidence that the two major international conferences, Vienna on Human Rights (1993) and Cairo on population (1994), coming so closely on the heels of Walker's efforts, finally figured FGM as a major humanitarian challenge.

Yet, in the United States, where girls of African origin are clearly at risk, problems in reception of Walker's book and film surfaced immediately. Now, Walker knew to be fearful. In *Anything We Love Can Be Saved. A Writer's Activism*, she records her apprehension. Visiting Jung's home in Bollingen, the final quaff of inspiration taken, she notes: "This was the last journey I had to make before beginning ... *Possessing the Secret of Joy*, a story whose subject frankly frightened me. An unpopular story. Even a taboo one" (126).

Although the USA gave the novel mixed reviews at best, the film incurred outright hostility. Led by African women intellectuals residing in North America, voices of resentment against Walker's violation of boundaries resonated loudly. *Newsweek*, for instance, quoted Sudan's premier female surgeon, Nahid Toubia, alleging that only because Walker's popularity had suffered did she take on FGM; a „falling star," the author was trying to get „the limelight" back (Levin "Alice" 244). More to the point, Walker, it was said, failed at empathy, her accusatory finger not illuminating but condemning and, therefore, insulting the audience she claimed to address, African women perpetrator-victims.

How affected Walker was by her critics can be teased out of the speech that opened *Warrior Marks'* tenth screening on February 24, 1994, in Oakland. "What can *you* do?" she asks her opponents. "... Refrain from spending more than ten minutes stoning or attempting to malign the messenger. Within those minutes thousands of children will be mutilated. Your idle words will have the rumble of muffled screams beneath them" (Heaven 63). Yes, she concedes, victims "will have to stand up for themselves, and ... put an end to it. But that they need our help is indisputable" (63). Indeed.

Yet, ironically, Walker's compassion is siphoned from a pool of shared African-*American* suffering, "our centuries-long insecurity" (Anything 150). This presumption of solidarity, not with former slaves but with immigrants to the United States, proves the lightning rod to her African critics. She dares them to know "who we are [and] ... what we've done to ourselves in the name of religion, male domination, female shame or terrible ignorance" (150). But who are 'we'? An African-American assumes commonality with African immigrants in the USA -- an unreciprocated move.

What is unbearable to Walker's critics, but incontrovertibly central to events, is named by Verena Stefan in a chapter devoted to Tashi in *Rauh, wild & frei: Mädchengestalten in der Literatur* [Rough, Tough and Free: Images of Girls in Literature]. Stefan reads the murder of the Tsunga (Walker's invented term for the exciseuse in *Possessing the Secret of Joy*) as a dagger to the shibboleth, reverence for the matriarch, in part inculcated by an authoritarian culture.[53]   Stefan reads the broken trust of FGM as "betrayal of girls by their mothers," and critiques the "control that the older wield over the younger" (108). Furthermore, in a revealing contrast, Stefan takes the rapport between Celie and Shug in *The Color Purple* as the other side of clitoridectomy's perfidy: "During circumcision, we have shared intimacy between woman and girl, but it is the intimacy of horror. Women observe and touch a young girl's genital – to mutilate it" (108).

In *The Dynamics of African Feminism. Defining and Classifying African Feminist Literatures*, Susan Arndt confronts homophobia, fear of which propels Stefan's observations, and also, I believe, accounts in part for Walker's negative U.S. reception. In particular, two theorists of African womanism – Mary E. Modupe Kolawole and Chikwenye Okonjo Ogunyemi – frankly come out as homophobes.

---

53 See Dympna Ugwu-Oju. *What will my Mother say? A tribal African girl comes of age in America*. Chicago: Bonus Books, 1995. Although African-American parents tend to be stricter than those of other U.S. minorities, their authoritarian child-rearing pales beside portrayals of parent-child interaction in Ugwu-Oju or in Sembene's *Moolaadé*.

Arndt writes, with inappropriate neutrality: "I do not know of any feminist theoretician in the West who dissociates him- or herself explicitly from lesbianism.... [So] it is a novelty within the feminist discourse that theoreticians of gender issues like Ogunyemi and Kolawole reject lesbian love explicitly, generally and firmly" (53).

Now, as most of us know, homophobia kills – and overcoming it is one of the last frontiers. It was homophobia that made 'offensive' to some critics *Possessing the Secret of Joy*; and the lesbian subtext is doubtless present in *Warrior Marks*. Pratibha Parmar, after all, lives an openly lesbian lifestyle. Although "a film maker first and last," and not a "lesbian filmmaker," Parmar has, for instance, in her film "'Jodie: An Icon' ... looked at ways in which ... Foster has been constructed ... for lesbians in her various screen personas" (Lola 38). Juxtaposed with a programmatic statement by novelist Buchi Emecheta, the conflict becomes clear. Emecheta charges Western feminists with concern only for "issues... relevant to themselves ... transplant[ed] onto Africa. Their own preoccupations – female sexuality, lesbianism and female circumcision – are not [African women's] priorities" (Ravell-Pinto 50).

I disagree. Consider Fanny Ann Eddy, a thirty-year-old human rights activist, who, in 2002, had founded a Lesbian and Gay Association in Sierra Leone. On September 29, 2004, Eddy was murdered, having earlier told the Human Rights Commission of the United Nations in Geneva how dangerous it was for lesbians, gays, bi- and trans-sexuals to remain invisible in African society (Guido 13).

Invisible, too, had been FGM. But breaking taboo has its price, and Walker suffered for her courage. Was she still ill at ease with the topic more than a decade later? Would her retirement be permanent? I asked Efua Dorkenoo, OBE, advisor and participant in *Warrior Marks,* if she knew why Alice had stopped campaigning. After all, I thought, Walker intervenes for other, important, abuses and "insists upon the necessary and strong relationship between spirituality, activism and art" (Griffin 23). Mainly, Efua told me, Alice had not forgotten the broad misrepresentation of her motives, the charges of ignorance of Africa and of self-interest. "That hurt her quite a lot, leading to the decision [to step back]. Her aim had always been to use her talent as an artist to bring the subject to the world. Having done that, maybe she feels justified in moving on to other things" (interview).

Typical of critics who misunderstand her aims, Jo Ellen Fair sees Walker's presence in her documentary as "paternalism," giving it an aura of "Westerners know best" (12). But *Warrior Marks* is a mixed-genre, as much a portrait of the artist-activist as a plea for girls. The "Amt für Frauen" [Women's Buro] in Schöneberg, Berlin, agrees, praising the documentary's "poetic narrative." Like a *bildungsroman,* "without recourse to bloody sensational scenes, [it] gets under your skin" (Amt für Frauen). Behind Fair's judgment, in contrast, is an unconscious but tenacious a priori: only Africans own this issue. Awa Thiam, for one, firmly disagrees. When asked, "How do you feel about us coming here and making this film?" she tells Pratibha Parmar: "'You know, I work ... in the belief [in] universal sisterhood, that we are all in this together" (Lolapress 36).[54]

Sadly, where allies should be, they often aren't. Take, for instance, Yari Yari Pamberi, that remarkable gathering at NYU in October 2004, and Walker's presentation at the end of a star-studded panel. Not a word about FGM, this omission of a piece with Ousmane Sembene's *Moolaadé* having been on the conference schedule but taken off. Lincoln Center provided a better venue. To his credit, as he announced the venue change, host and organizer Manthia Diawara gave a few militant words to FGM. But as recently as 2004, the elite of the African and African-American intelligentsia continued tight-lipped over torture.[55]

How different has been reception in Europe, particularly in Germany, where more than 20,000 African girls are at risk. Walker's input has not only been welcomed but acted upon. For instance, following a major interpellation in 1998, Walker is named by the Bundestag as an inspiration in asylum debates.[56] The decision to offer sanctuary to African women fleeing the threat of excision is, in part, based on Walker's work. "In many respects, genital mutilation

---

54 Pratibha answers, "It was refreshing for me to hear her, having come from the late 80s and early 90s talk of post-feminism, with so much cynicism around the bankruptcy of feminism. It was good to meet a woman who still has that wonderfully optimistic belief in the idea of universal sisterhood, which many of us held in the 70s, but which got lost along the way as we got more fragmented and disparate" (Lolapress 36-37).

55 Another distressing but typical example: Ada Uzuamaka Azodo, in „Issues in African Feminism: A Syllabus," includes nothing on FGM – in 1997! The course was offered at Indiana University Northwest.

56 See http://www.ak-kipro.de/fgm_bundestag200107.html. Accessed 17 July 2004

resembles torture and violates the human right to bodily integrity," the Bundestag concurs (Drucksache 14/5285). And the document goes on: „Human rights organizations like Terre des Femmes and human rights activists like Alice Walker have been pushing the theme into public awareness for years." Footnoted is "Walker, Alice/Parmar, Pratibha, *Narben oder die Beschneidung der weiblichen Sexualität*, Hamburg 1996," the German translation of *Warrior Marks*. What's more, at a preparatory multi-partisan hearing organized by the Green Party in Bonn, Dr. Angelika Köster-Lossack, MdB (Member of Parliament) embedded the German title of Walker's film -- *Narben [Scars]* -- in the title of her talk. Terres des Femmes, the German NGO advocating for women's human rights commended by the Bundestag, took its motto directly from Walker: "Resistance is the secret of joy," the closing lines of *Possessing the Secret of Joy.* Alice Schwarzer, Germany's ‚first' feminist and editor of *EMMA*, has featured Walker and *Narben* in her magazine's pages. In 1996, Christa Müller, wife of Oskar Lafontaine, former head of the Social Democratic Party, advertised *Narben* on the program of her association's inaugural conference, the keynote address delivered by FORWARD'S Comfort Ottah who appears in Walker's film. In fact, *Warrior Marks* has been distributed throughout the nation; nearly every major university and *gymnasium* has it in its archives or has shown it at film festivals and other events (Levin "Alice" 246).

This astonishing diffusion and approval have not been due to suppression of resident African women's voices, for their opinions have been sought. They are engaged in many of the NGOs feeding into governmental agencies concerned with refugees, foreign aid, and women. Finally, government has accepted the responsibility of protection, even if there remains a great deal to be done toward implementation. Indeed, a network for all groups working on FGM within Germany and abroad, INTEGRA,[57] held its inaugural conference in Berlin on December 12-13, 2006. It acts under the patronage of none other than the President of Germany, at that time Dr. Horst Köhler.

In conclusion, what has made such a difference in reception? Why should Alice Walker be shunned in the USA for dealing with this issue but celebrated in Europe, especially in Germany? Reasons are, of course, complex, and have to do with the development of the women's movement on both continents; on what has been learned, in Germany, from the Holocaust; on what is permitted in terms of who may speak „for" whom, and on the preference for human rights discourse in Europe as opposed to post-structuralist shattering of needed coalitions in the USA with its stubborn, divisive, essentialist stance on "race." Though no one will argue the absence of racism in Europe, the history of its

---

57 To maintain his association's independence, Rüdiger Nehberg refused to enroll Target e.V. as a member of INTEGRA that now claims 20 members. Target, however, has made headlines for a conference it sponsored on 22-23 November 2006 in Cairo, chaired by the Grand Mufti of Al Azhar University, Professor Ali Goma'a. Nehberg and his associate Annette Weber showed imams from throughout the Islamic world an explicit film of the operation. The result was issuance of a Fatwa: "FGM is a crime that violates Islam's most cherished principles." (See "Karawane der Hoffnung," 2006).

institutionalization is very, very different – and has led to a striking distinction in Alice Walker's influence on movements against FGM.

Returning to the Rose

Can it be a coincidence that Walker's most famous essay is titled "In Search of Our Mother's Gardens"? In it, she confronts the suppression of creativity in the formerly enslaved and their impoverished female descendants who, like Walker's mother, lacking canvas and oils, poured their souls into beauty afforded by nature – they cultivated gardens. Flora play an inordinate role in Walker's work. *The Color Purple* adorns the fields. The divine embodied in a wild flower, it dare not be overlooked, for "anything we love can be saved."

Happily, this promise even covers amputated organs, for a pioneer in reconstructive surgery can now restore the clitoris. Pierre Foldès, a urologist, has developed a procedure that consists of trimming the scar from the clitoral surface before rolling back tissue along the root to expose the still-existing organ, for the clitoris extends backward from its tip to the 3rd and 4th vertebrae. It cannot, therefore, be truly excised, and more than 75% of patients regain some sensitivity.

He reunites the cut rose with its root.

WORKS CITED

Abdi, Maryam Sheikh. "The Cut." 2006.
http://www.popcouncil.org/rh/thecut.html

Accessed 3 February 2007. For permission to reprint, contact
pubinfo@popcouncil.org

Abdi, Nura and Leo G. Linder. *Tränen im Sand*. Bergisch Gladbach: Verlagsgruppe Lübbe, 2003. Excerpted as "Watering the Dunes with Tears." In *Feminist Europa. Review of Books.*28-33.
http://www.ddv-verlag.de/issn_1570_0038_jahr2003.4_vol01_nr01.pdf

Accessed 2 February 2007.

Abuscharaf, Rogaia Mustafa, ed. *Female Circumcision: Multi-cultural Perspectives*. Philadelphia: U. of Pennsylvania P., 2006.

Alvaro, Alexander, et al, eds. *DON'T. Women's Art Protest against Female Genital Mutilation*. Lünen, Germany: Drückerei Peter Holtkamp GmbH, 2006.

Amt für Frauen. Bezirksamt Schöneberg von Berlin. Flyer distributed at a showing of *Warrior Marks*. 28 May 1997. Rathaus Schöneberg.

Angier, Natalie. *Woman. An Intimate Geography.* NY: Anchor, 2000.

Antwort der Bundesregierung. Drucksache 14/5285. http://www.ak-kipro.de/fgm_bundestag200107.html

Accessed 5 April 2005.

Arndt, Susan. *The Dynamics of African Feminism. Defining and Classifying African Feminist Literatures.* Trenton: Africa World Press, 2002.

Azodo, Ada Uzuamaka. „Issues in African Feminism: A Syllabus." *Women's Studies Quarterly. Teaching African Literatures in a Global Economy.* Vol. XXV, Nos. 3 & 4. Fall/Winter 1997. 201-207.

Blake, William. "The Sick Rose." http://www.artofeurope.com/black /bla1.html

Accessed 1 January 2007.

Dorkenoo, Efua. *Cutting the Rose. Female Genital Mutilation. The Practice and Its Prevention.* London: Minority Rights Group, 1994.

_____. Telephone interview. 18 April 2005.

Douglas, Mary. *Purity and Danger. An analysis of concept [sic] of pollution and taboo.* NY: Routledge, 2005. (First published 1966).

Dudones, Jill. "The Unkindest Cut." *Ms. Magazine.* Winter 2007. 29.

Fair, Jo Ellen. "The Lives of Women in Africa." *Feminist Collections. A Quarterly of Women's Studies Resources.* Vol. 19. No. 4, Summer 1998. 11-13.

Griffin, Farah Jasmine. "The courage of her convictions." Rev. of Alice Walker. *Anything We Love Can Be Saved: A Writer's Activism.* NY: Random House, 1997. *The Women's Review of Books.* Vol. XV, No. 4/ January 1998. 23-24.

Gruber, Franziska. "Aktionstag gegen Genitalverstümmelung." *Terres des Femmes – Menschenrechte für die Frau.* 2/2006. 12-13.

Guido, Anke. "Menschenrechtsverletzungen an lesbischen Frauen." *TdF Menschenrechte für die Frau.* 2/2005. 12-13.

Nehberg, Rüdiger and Annette Weber. *Karawane der Hoffnung. FGM Anti-Islam.* http://www.ruediger-nehberg.de/
Accessed 01.12.2006.

Korn, Fadumo and Sabine Eichhorst. *Born in the Big Rains. A Memoir of Somalia and Survival*. Trans. and Afterword. Tobe Levin. NY: The Feminist Press, 2006.

Levin, Tobe. "Alice Walker: Matron of FORWARD." *Black Imagination and the Middle Passage*. Eds. Maria Diedrich, Henry Louis Gates Jr. and Carl Pedersen. NY: Oxford UP, 1999. 240-254.

_____. „Female Genital Mutilation and Human Rights." *Comparative American Studies. An International Journal*. Special issue on human rights. Guest eds. Werner Sollors and Winfried Fluck. Vol. 1 (3) 2003. 285-316.

_____. „Women as Scapegoats of Culture and Cult: An Activist's View of Female Circumcision in Thiong'o's *The River Between*." *Ngambika. Studies of Women in African Literature*. Ed. Carole Boyce Davies and Anne Adams Graves. Trenton, NJ: Africa World P., 1985. 205-221.

Lolapress Europe. "Identities, Passions and Commitments. An interview with the British Filmmaker Pratibha Parmar." *Lola Press. International Feminist Magazine*. Nov. 99 – April 2000. No. 12. 36-41.

Parmar, Pratibha, dir. *Warrior Marks*. Prod. Alice Walker. Distr. Women Make Movies, 1993.

Parrot, Andrea and Nina Cummings. *Forsaken Females. The Global Brutalization of Women*. NY: Rowman & Littlefield, 2006.

Prolongeau, Hubert. *Victoire sur l'Excision. Pierre Foldès, le chirurgien qui redonne l'espoir aux femmes mutilées*. Préface de Bernard Kouchner. Paris: Éditions Albin Michel, 2006.

Prolongeau, Hubert. *Undoing FGM. Pierre Foldes, the Surgeon Who Restores the Clitoris*. Foreword Bernard Kouchner. Trans. and Afterword Tobe Levin. Frankfurt am Main: UnCUT/VOICES Press, 2011.

Ravell-Pinto, Thelma. „Buchi Emecheta at Spelman College." *SAGE. A Scholarly Journal on Black Women*. 2-1 (Spring 1985): 50-51.

Reinharz, Shulamith, Tobe Levin and Joy Keshi Walker, eds. *Through the Eyes of Nigerian Artists. Confronting Female Genital Mutilation*. Frankfurt: Mainprint, 2006.

Sembène, Ousmane, dir. *Moolaadé*. Prod. Filmi Doomireew (Senegal), 2002. New York Videos, 2004.

Stefan, Verena. *Rauh, wild & frei: Mädchengestalten in der Literatur*. Frankfurt/Main: Fischer, 1997.

Thiong'o, Ngugi wa. *The River Between*. London: Heinemann, 1956.

Ugwu-Oju, Dympna. *What will my Mother say? A tribal African girl comes of age in America*. Chicago: Bonus Books, 1995.

Walker, Alice. "Heaven Belongs to You." *Re-visioning Feminism around the World*. NY: The Feminist Press, 1995. 62-63.

_____. "Heaven Belongs to You. *Warrior Marks* as a Liberation Film." *Anything We Love Can Be Saved. A Writer's Activism*. NY: Random House, 1997. 147-151.

_____. "In Search of Our Mothers' Gardens." *In Search of Our Mothers' Gardens: Womanist Prose*. NY: Harcourt, 1983.

_____. *Possessing the Secret of Joy*. NY: Harcourt, Brace Jovanovich, 1992.

_____. (with Pratibha Parmar). *Warrior Marks: Female Genital Mutilation and the Sexual Blinding of Women*. NY: Harcourt, 1993.

3

Rose Symbolism in Qabalistic Tarot and Beyond

By Albert Amao

The Qabalistic Tarot[58], a pictorial collection of seventy-eight Arcana or cards, is divided into twenty-two Major Arcana, the most symbolic keys of the Tarot system, and fifty-six Minor Arcana. For the purpose of this essay, discussion will be limited to the twenty-two Major Arcana, numbered 0 to 21. Here we survey the symbolism of the rose in the twenty-two major Arcana of the Tarot tradition and offer an explanation as to why the rose is an important symbol in the Rosicrucian Brotherhood. A second focus of the essay is the Qabalistic Tree of Life, which is a major component in metaphysical teachings.

The symbol of the rose appears seven times in the Major Arcana, which have often been considered archetypes

---

58   The term 'Qabalah' is perhaps most recognized when spelled 'Kaballah.' However, in ancient Hebrew there was no letter 'K'. Therefore, for the purpose of this essay, the ancient spelling of this teaching will be used. This essay is an adapted and expanded version of a subchapter of my book entitled *Aquarian Age and the Andean Prophecy* (Indiana: AuthorHouse, 2006).

to explore inner realms.[59] The rose's seven appearances in this system are not fortuitous. Seven has often been considered a sacred number in Holy Scriptures. The Christian *Bible* speaks of 'seven deadly sins,' and the symbolic 'seven days of creation.' According to the book of *Genesis*, God created the universe in seven days. However, today it is accepted by many philosophers, historians and recently by major theologians, such as John Shelby Spong, a retired Episcopal Bishop of Newark New Jersey, that even the Christian *Bible* used symbolic writing at times. Spong states, **"Belief in the historic accuracy of these texts no longer exists in academic circles..."**[60] Though the Christian *Bible* recounts that the world was created in seven days, the story should not be translated literally. And yet, the Christian *Bible* is not the only place in which the number seven appears. In nature, the rainbow has seven colors. There are seven days in the week, and seven chakras or hidden energy centers within the human body. As the reoccurrence of the number seven in seemingly familiar aspects of daily life is not purely coincidental, neither are the seven appearances of the rose in the Major Arcana of the Tarot system.

The Qabalistic Tarot tradition is based on Qabalah, the esoteric interpretation of the Jewish Scriptures.[61] However

---

59   See Paul A. Clark, *TAROT: The Magical Keys to Consciousness.* Paul Clark is the founder of the metaphysical school: Fraternity of the Hidden Light.

60   John Shelby Spong, *BORN OF A WOMAN: A Bishop Rethinks the Virgin Birth and the Treatment of Women by a Male-Dominated Church* (San Francisco: Harper Collins Publishers, 1992) 3.

61   See *The Columbia Electronic Encyclopedia,* Sixth Edition, (Columbia University Press, 2003).

modern scholar Lewis Keizer, author of *The Esoteric Origins of Tarot: More than a Wicked Pack of Cards*, refutes this claim, stating that the modern esoteric tarot decks are a development stemming from the medieval Italian Tarocchi, a pastime that was popular throughout Medieval Italy where the images were often interpreted in magical games.[62] Throughout this essay, the term 'Tarot Keys' is used, instead of 'Tarot Cards,' in order to differentiate from the incorrect notion that the Tarot system is a means to foresee the future.[63] Each key of the Tarot system encapsulates a great deal of wisdom; thus the name Arcana, which the *American Heritage Dictionary* defines as "Specialized knowledge or detail that is mysterious to the average person."

There is a secret and subtle language in the symbols of the Tarot configuration which speaks directly to the subconscious mind. Few are aware of this arcane language because it is not traditionally taught on any level throughout the academy. Dr. Robert Wang, author of *The Qabalistic Tarot,* comments on the lack of focus that academic circles have given to the imagery of the Esoteric Tarot Keys, which he claims could be used as a tool for personal self-exploration. Wang explicates, "Despite the increased public interest, surprisingly little attention has been paid to the Tarot by academia, though the cards are a veritable gold mine of art

---

62   For further information on the origins of the Esoteric Tarot see *The Esoteric Origins of Tarot: More than a Wicked Pack of Cards* by Lewis Keizer, PhD, and Chapter II of *Beyond Conventional Wisdom*.

63   Refer to Albert Amao's *Beyond Conventional Wisdom* (Indiana: AuthorHouse 2006).

history and metaphysical philosophy." [64] However, everyone is surrounded by symbols in myriad forms, as the symbols are part of daily life. In Freemasonry fraternities as well as metaphysical and religious organizations, symbols are presented as emblems of identification within institutions that share the same ideological or religious principles. Many of the symbols taken for granted are conventional signs of religious, mystical, or metaphysical organizations when in fact they are a subtle language to convey an invisible dimension that cannot be expressed in conventional words.[65] Dr. Cynthia Giles, Jungian Psychologist explains the importance of symbols and consciousness-raising: "These multiple levels of the archetypes are all present in the Tarot images and it is for this reason that the Tarot works in a variety of ways. For the Tarot reader images are the point where a process of archetypal analysis begins, and from the images unfold the behaviors and styles of consciousness."[66]

Over time, the symbols of the Tarot Keys have acquired layers of increasingly complex meaning. This evolution tells a great deal about perception of the inner world and how one perceives the nature of life and of the universe. Furthermore, each Tarot Key constitutes a single entity that can unlock wisdom which already exists in the mind of the observer. The combination of one Tarot Key with another Key can

---

64   Robert Wang, *The Qabalistic Tarot* (Samuel Weiser, Inc., 1983) 2.
65   For more information about the special language of symbols refer to the third chapter of the book *Aquarian Age and the Andean Prophecy* mentioned above.
66   Cynthia Giles, *The Tarot, History, Mystery, and Lore,* (A Fireside Book, 1994) 60.

bring to mind a wealth of knowledge for those who know how to properly view the images. That is why the sages of this wisdom, such as Paul Foster Case, perhaps one of the greatest Qabalists and Tarotists of the twentieth century, has admonished in his book, *The Tarot: A Key to the Wisdom of the Ages*, that a man cannot exhaust all the knowledge summarized in this Qabalah in a single lifetime.

Tarot key 0, The Fool (BOTA)

Explaining the symbolism of Tarot will begin with the first Tarot Key, numbered 0, and often called The Fool. In this key, one can observe a young man at the top of a mountain, clutching a white rose in his left hand. Yet, he does so without pricking his fingers on the thorns of the rose stem. Since pricking one's fingers on the thorns of a rose would generally cause a certain level of pain and suffering, the fact that the chap grasps the rose unaffected by the thorns would suggest that he is free from any terrestrial affliction. The meaning of the white rose signifies purity and innocence.[67] The general meaning of a rose is *desire*; in this case, a white rose conveys the idea of innocence, purity and higher aspirations. The white rose gives the basic idea to this key which represents the Universal Life-force, the energy that is behind growth and development. This is reinforced by the gesture of the young man looking upward to the heavens, toward higher aspirations. Roses, along with another flower, the lily, can be observed in other Tarot Keys as well, such as the Key of the Magician.

67 Hajo Banzhaf, *Tarot for Everyone* (Urania, Switzerland: AGM AGMuller, 2005) 32. Anthony Louis, *Tarot: Plain and Simple* (St. Paul Minnesota: Llewellyn Publications, 2004) 52.

Tarot Key 1, The Magician (BOTA)

In the Tarot Key numbered 1, a Magician stands in a garden of red roses and white lilies. There are also roses hanging above the Magician's head, insinuating a rudimentary shelter. In this key, the color of the roses is red, in contrast to the white rose on the Fool's key. The meaning behind each color is different; the red rose signifies active desire. The white lilies in the garden indicate knowledge, higher aspirations, abstract truth, and purity of the soul. The Magician is looking at the table, which represents the level of reality in which the Magician exits.[68] Displayed on the table are four items, a wand, a sword, a cup, and a coin. The wand symbolizes will, the sword represents action and ideas, the cup signifies imagination and emotions, and the coin is associated with material things. Each symbol suggests a task or element of daily life that one must master. Knowingly or unknowingly, every human being is a Magician by virtue, and possesses the potential to create his own reality through the power of his attention, thoughts and emotions. One would be a white magician if he has positive intentions and does good deeds for his fellow neighbors. One would be a black magician if he fosters negative emotions and tries to take advantage of his fellow neighbors to satisfy his egotistical needs.[69]

Contrary to Four Noble Truths of the Buddhist doctrine, where it is suggested that man should suppress desire to avoid suffering, Qabalistic doctrine considers desires to be blessings.[70] In Buddhist doctrine, desire of material possessions should be suppressed while desire for enlightenment should be fostered.   In Qabalah, it is said that God leads man toward his liberation *through* his desires. In fact, desire is the motivational force that impels man to action. That is why masters of wisdom, such as Buddha himself, recommended

---

68   Ibid. 34.
69   Rev. Ann Davies, *Inspirational Thoughts on the Tarot* (Builders of the Adytum) 8.
70   According to Buddhist teaching, the *Four Noble Truths* are: 1. Suffering exists; 2. Suffering arises from attachment to desires; 3. Suffering ceases when attachment to desire ceases; 4. Freedom from suffering is possible by practicing the Eightfold Path. For more information regarding the Four Noble Truths, see: "Four Noble Truths." *Britannica Concise Encyclopedia*, Encyclopedia Britannica, Inc., 2006.

that their pupils have a steady desire for enlightenment. It is that strong desire that would allow the pupil to endure all the efforts and sacrifices that are necessary to reach that end objective. Man without desire is a virtually dead person; he would not have any reason to live purposefully. Essentially, the Magician represents one's craving to self-improve. Like the Magician, man has all the necessary tools before him, but it is his *will*, or desire that drives him towards action.

In the Tarot Key 1, there is something regarding the roses of the Magician's garden that is worthy of a note. Roses have five petals, or some multiple of five petals; thus roses represent the Pentagram, the five-pointed star displayed on the Magician's table, as well as man, the microcosm. On the other hand, the lilies have six petals, and therefore stand for the hexagram, the macrocosm, that is the universe.[71] If one puts together these two digits, it creates the algorithm sixty-five. And sixty-five is the numerical value of the Latin word *LVX*[72] (50+5+10 = 65), which translates to the word 'Light' in English. Furthermore, sixty-five is also the numerical value of the Hebrew word *Adonai*, which translates to the divine word 'Lord.'[73]

The next Tarot Key in which the rose is present is Key 3, The Empress. The fundamental idea underlining this Tarot Key is very similar to the esoteric meaning behind the

---

71   Paul F. Case, *The Tarot, A Key to the Wisdom of the Ages*, (BOTA, 1990) 45. Éliphas Levi, *Transcendental Magic, Its Doctrine & Ritual* (London: Braken Books, 1995) 89-98.
72   Classical Latin had no letter 'U', so the letter 'V' would be in place of the letter 'U'.
73   Israel Regardie, *Foundations of Practical Magic* (The Aquarian Press, 1979) 127.

rose. The whole idea is that The Empress *typifies* Venus, the *desire nature* expressed through creative *imagination*. "She embodies creative processes and lively growth and the birth of the new."[74] This key is associated with the planet Venus. In the Rider/Waite version of the image, The Empress has the glyph of Venus on the shield which is at her right side, and her entire garment reflects the Symbol of Venus.

Key 3, The Empress (BOTA)

Looking closely at the Empress, one can see that she is pregnant. In fact her sovereignty as Empress is based on her reproductive power, which is the basis for the existence of

74   Hajo Banzhaf, *Tarot for Everyone*. 38.

life and for the flowing of the Life-force into new forms. This idea is reinforced by the existence of the stream and waterfall present behind the Empress. The stream symbolizes the water of life, and the waterfall suggests the idea of union of male and female. The male is symbolized by the stream flowing into the pool beneath it, which represents the female. This is a beautiful pictogram chosen to represent the reproductive power of Mother Nature and the flowing of the Life-energy! The Empress in Tarot Key 3 represents Mother Nature. If there is no fecundation, there is no reproduction, and consequently, there is no life. The reproductive power of the Empress is demonstrated by the fact that she holds a scepter, a symbol of power and sovereignty. The scepter is resting on her womb, the vehicle for reproduction. The basic meaning of this key is duplication, multiplication, and reproduction in all areas of life.[75] Dr. Arthur Edward Waite, a scholar on Rosicrucianism, Qabalah and Tarot, has offered an explanation of this Tarot key by stating, "The card of the Empress signifies the door or gate by which an entrance is obtained into this life, as into the garden of Venus."[76] On the right side of the Empress there are five red roses representing the five senses. The five physical senses incite desire, in other words, every desire is rooted in the physical senses. As pointed out before, the rose represents the Pentagram, and the Pentagram is also a representation of man. Based on the teaching of the principal of karma, it

---

75　Ibid.
76　Dr. Arthur Edward Waite, *The Pictorial key to the Tarot* (U.S. Games System, Inc, 1989) 83.

can be said that man was incarnated on earth by his own intense desire to come to this material plane in order to learn and expand his consciousness through physical experiences in life.[77]

The next key that contains the rose is the Key 5. In this key, two priests are keeling in front of the High Priest, known as the Hierophant. The priest on the left is wearing a robe decorated with roses; the priest on the right is wearing a robe adorned with lilies.[78] These are the same flowers that are part of the magician's garden. Thus they stand for desire and knowledge. The Hierophant is the revealer and teacher of sacred mysteries. Only those ready to listen to their Inner Voice and who adopt a humble attitude will benefit from the mysteries revealed by the High Priest.[79]

---

77   See *A Dictionary of Buddhism* (Oxford University Press, 2004).
78   To fully appreciate the color pictures, it is recommended that one procure the book of Paul F. Case, *The Tarot: A Key to the Wisdom of the Ages.*
79   Hajo Banzhaf, *Tarot for Everyone.* 42.

Key 5, The Hierophant (BOTA)

In Tarot Key 6, the symbol of the rose is implicit. Key 6 entitled 'The Lovers,' depicts the Christian Biblical Garden of Eden. The key shows Adam, Eve and the so called Serpent of Temptation. Behind the woman is the Tree of Knowledge of Good and Evil, and behind the man is the Tree of Life.[80] The five fruits on the Tree of Knowledge of Good and Evil are the occult representation of the five red roses of the Magician's garden, thus, they stand for the five senses. The five senses are the tempters of man, which is why the Serpent of Temptation is coiling up on the tree behind Eve. It is interesting to note that the Christian *Bible* never mentions the apple in the story

80   Ibid. 44.

of Adam and Eve. However, today it is a common held belief that the fruit Eve was tempted by was the apple. Even the *Catholic Encyclopedia* mentions that Eve's fatal mistake was eating the apple.[81] It should be noted, however, that when one cuts an apple horizontally down the middle, one can see the Pentagram in each side of the apple, with one seed inside each arm of the Pentagram. Is it not marvelous? The Pentagram epitomizes man, the microcosm, and appears in the very piece of nature that was the alleged tool that caused the downfall of man. Continuing with the allegory of Adam and Eve, the Devil allegedly told Eve that the fruit was ripened and therefore could be eaten. She was instructed that when she ate from the fruit her eyes would be opened and she would be like God, knowing good and evil.[82] Interestingly enough, the *esoteric* elucidation of this story reveals that the Tempter and the Redeemer are the same. At times, the devil is also called 'The Deceiver,' 'The Father of Lies,' or 'The Master of Appearances.' In fact, the devil deceives man through man's own five physical senses and with the appearances presented in the world.[83]

The next key where the roses appear is in Key 8, Strength. The roses are presented as a chain that encircles the waist of the lady and the neck of a lion that stands beside her. As previously mentioned, roses symbolize desire. Hence, the

---

81   See *Catholic Encyclopedia*, s.v. 'Sext.'
82   Genesis, Chapter 3.
83   This assertion has been extensively discussed in the essay, "The Adversary and the Redeemer," which can be found in the book *Beyond Conventional Wisdom* by Albert Amao.

chain of roses is a series of desires woven together. Rightly cultivated, desires are the most potent forms of suggestion. The young lady presented in this key represents the purified subconscious mind that has dominion over nature through creative imagination; she exerts control over the forces below the human level of manifestation, represented by the lion. She is wearing a white robe, which represents purity, wisdom, light, and knowledge.[84] The meaning behind this key is that the subconscious has mastered natural forces through the law of suggestion. When desires are purified, the subconscious, which has perfect power over the lion, will become an ally of man instead of his enemy. In the Tarot doctrine, the lion, king of nature's animal kingdom, represents all subhuman forces of the cosmic energy. Key 8 depicts the lion being under the control of the lady clothed in a white robe.[85]

The final representation of the rose in the Tarot's Major Arcana is present in Key 13, Death. The first impression of this key is demise, desolation, and darkness. But occult awareness states that there is no death; there is only a process of transformation and change. The natural Law of Transformation brings about dissolution and change.[86] Upon careful observation of this key, one can see the symbol of a seed in the upper left corner and a white rose in the middle of the right side of the key. The white rose is that of Key 0, and has the same meaning as the rose in the hand of the

---

84 Banzhaf, Ibid. 54.
85 Ibid.
86 Refer to the chapter, "The Dialectic and the Hermetic Philosophy" in *Beyond Conventional Wisdom* by Albert Amao.

Fool, that is purity and innocence. Therefore, Death implies rebirth; this is insinuated by the fact that the yellow sun in the background of Key 13 is rising in the East. The river, an emblem of the flow of life, is making a bend directing its current to the sun. The seed in the upper left corner has five rays and symbolizes rebirth and life.[87]

Key 13, Death (BOTA)

Ultimately, this Tarot key symbolizes the ending of the Piscean Age, a time characterized by global religion, and the renaissance of humanity into a new era, the Aquarian Age, an age of people being in an advanced stage of spiritual consciousness. This is insinuated in Tarot Key 13 by the fact

87   Banzhaf, Ibid. 58.

that the key shows only one foot, indicating the end of the Piscean Age. In Astrology, signs have their correspondence to different parts of the human body; the sign of Pisces corresponds to the feet.[88] Along with Tarot, the rose has been associated with other spiritual circles. When the symbol of the rose is associated with religious imagery, it has a special spiritual dimension. Meditation on it produces a profound effect in the psyche of a person. An example of this is the Christian symbol of the 'Rosy Cross.'[89]

The symbol of the 'Rosy Cross,' also known as 'Rose Cross,' is a familiar symbol of the Rosicrucians, an organization dedicated to the "systematic approach to the study of higher wisdom that empowers you to find the answers to your questions about the workings of the universe, the interconnectedness of all life, your higher purpose, and how it all fits together."[90] The Rosicrucian Brotherhood is a Christian sect founded in Europe in the fifteenth century according to some sources. But other sources say its origins are long before the birth of Christ in some part of North Africa, as noted in other chapters in this book. The brotherhood combined the emblem of the cross and the rose, and adopted this icon as the trademark of what started out as a secret society.[91] In fact, the

---

88   For further information refer to *Aquarian Age & the Andean Prophecy* by Albert Amao.
89   For a more complete diagram as well as further links to the meaning of the 'Rosy Cross' and the emblems shown, consult http://altreligion.about.com/library/glossary/symbols/bldefsrosecross.html
90   For more information on the Rosicrucian organization, visit their official website, http://www.rosicrucian.org/home.html
91   Mark O'Connell and Raje Airey, *Signs and Symbols* (Hermes House), 175. Also see *Catholic Encyclopedia*, s.v. 'Rosicrucians.'

cross and the rose as separate emblems are two of the most ancient and universal symbols in human history. The original **'Rosy Cross,'** formulated as a Christian symbol in the second century, **was a** red rose positioned in the center of a cross.[92] This living image has both a mystical and esoteric meaning.

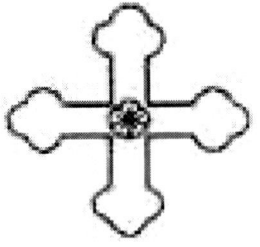   

Symbol of a Cross   Five Petals Rose

17th Century Rose Cross

92   Ibid. 108.

Different Illustrations of the Rosy Cross

In the nineteenth century, the symbol of the Rosy Cross was adopted by the Order of the Golden Dawn, an organization formed in 1888 that devotes itself to the study of ancient wisdom.[93] The magicians of the Golden Dawn transformed the Rosy Cross into a more sophisticated and complex icon as they incorporated Qabalistic, alchemical, and astrological symbols. When observed closely, the Golden Dawn Rosy Cross contains a rosy-cross in its central point, representing the Absolute from which everything is manifested. The Golden Dawn Rose Cross expresses the meaning of the ineffable and unpronounceable word of the Tetragrammaton, the four letters used to represent the name of Yahweh, a name that must not be spoken aloud in the Jewish faith. Encircling the innermost point of the cross are three Hebrew letters, Aleph (א), Mem (מ), and Shin (ש), known as the mother letters in the Hebrew alphabet. In turn, the three mother letters are surrounded by the seven Hebrew double letters, and consecutively, the seven letters are encircled by the twelve simple letters to form a rose of twenty-two petals. The three mother letters correspond to the three outer planets, the seven double letters to the seven traditional planets,[94] and the twelve simple Hebrew letters correlate to the twelve zodiacal signs. In total, this makes

---

93   See "Order of the Golden Dawn," *The Concise Oxford Companion to Irish Literature* (Oxford University Press, 2003).
94   For convenience of this essay, the Moon and the Sun are considered as planets.

twenty-two, equivalent to the twenty-two Tarot Keys and the twenty-two Qabalistic Paths of the Tree of Life.

The Golden Dawn Rose Cross Lamen

The Golden Dawn Rosy Cross, Lamen, also embodies the Qabalisitc Tree of Life, a mystical concept of Qabalah which is used to understand the nature of God. The middle Pillar of the Tree is represented by the vertical cross-bar, which has three squares and the rose composed of Hebrew letters or petals occupying the center of the cross. The yellow square at the top of the vertical line is the *Sefirah Kether*, Crown of the Tree of Life. The rose petals represent the Sefirah of Tifaret, or Beauty. The Sefirah Yesod, or Foundation, has a glyph of the hexagram surrounded by the planetary symbols and the sign of the sun at its center. Finally, the bottom square of the vertical arm is the Sefirah Malkuth, or Kingdom. This is confirmed by the fact that this Sefirah is usually divided into four elements represented by the colors: citrine, olive, black and russet.

The other two columns of the Tree of Life are represented by the horizontal arms that have four triangles adjacent to them. In fact, the four triangles around the arms of the cross stand for the Sefirot: Mercy, Severity, Victory and Splendor. These rays bear the Latin letters that form the acronym I.N.R.I., meaning 'Jesus of Nazareth, the King of the Jews.' Each letter is associated with an astrological sign according to its Qabalistic correspondence. For instance: the letter I, 'Yod,' in Hebrew, corresponds to the astrological sign Virgo, the letter N, 'Nun,' to Scorpio, and the letter R, 'Resh,' is associated with the planet Sun.[95]

Another representation of the Golden Dawn Rose Cross Lamen depicts each arm of the cross with an upright pentagram containing the symbol of the four alchemical elements, Fire, Water, Air and Earth, and crowned with the symbol of the Spirit at the top of the pentagram. As mentioned, the pentagram, in esoteric geometry, stands for the dominion of the Spirit over the Elements. At the end of each arm of the cross there are three alchemical symbols representing: sulfur, salt, and mercury.

In Sacred Geometry, the cross is defined as the meeting of two lines, one horizontal and the other vertical. They

---

95　For more information on Kabbalah, visit the Kabbalah Centre's official website, http://www.kabbalah.com

are two perpendicular forces joined at a central point. The horizontal line represents the feminine principle and the vertical line represents the masculine principle. The cross is also an icon representing the physical world, epitomizing the dual nature of physical manifestation; therefore, it represents the physical expression of time and space. The vertical line is the concept of time while the horizontal line is the concept of space. The intersection of these two lines at a central point is an initial expression of the Life-force.[96] Incidentally, in Egyptian mythology, the crux Ansata, is similar to the Venus glyph, or symbol, ♀ . In astrology, this is the symbol of life, and as mentioned before, where it appeared in the Tarot Key of the pregnant Empress.

96 Ibid.

The Rose over a Calvary Cross

The Rosy Cross symbol is sometimes called a Calvary or Piscean Cross because the length of the vertical cross-bar is longer than the horizontal one. This was the characteristic of the cross during the Piscean Age. A rose is located over the intersection of the cross, encircled by eight thorns in the form of five-pointed stars. The cross is also surrounded by 'Ouroboros,' which is a serpent swallowing its tail. The 'Ouroboros' is an ancient universal symbol depicting time, continuity of life, wisdom, and eternity.[97] The tip of the serpent's tail equates to a point, and its head is a circle; hence, together, they make the astrological glyph of the sun, ☉ , that is, the point inside a circle. In the triangle over the cross there is the inscription of the Divine Name, Yod-Heh-Vav-Heh, in Hebrew characters, יהוה. As has been indicated above, one esoteric meaning of the cross is the material plane; the serpent in the form of a circle depicts the cycle of death and rebirth. Here the rose epitomizes the unfolding of the spiritual nature from the material one through pain and suffering.

Along with the cross, the symbol of the rose also has a deep mystical history that would later translate to religious significance as well. As previously mentioned, the white rose is the symbol of purity, desire, aspiration, virginity, truth, and wisdom. A red rose is an emblem of passion, fertility, life and death, and for a Christian analogy, represents the blood of

---

97   Mark O'Connell and Raje Airey, *Signs and Symbols* (Hermes House), 236.

Jesus Christ as well as the martyrs who died in his name.[98] The Rose Cross illustrates the triumph of spirit over matter, where the Rose, a symbol for love, is placed over the Cross, matter. The hidden significance of this symbol reveals the redemptive power of Jesus Christ's blood shed on the cross when, according to Christian tradition, he was crucified on the hill of Calvary.[99] Therefore, the message of the Christian Rosy Cross is redemption through a Savior.

As one can see from this essay, the symbol of the rose is not coincidental in the Tarot tradition. There is significant meaning behind each pictogram of the Tarot, especially the rose and its appearances in the Major Arcana. The significance behind such an emblem is carried through other esoteric circles, including the Rosicrucians, where the rose was adopted as their official symbol. Both the rose and its meaning cannot be overlooked when considering esoteric doctrine, for the symbol of this life force is present in everyone's daily life.

98 Ibid.175.
99 See *Catholic Encyclopedia*, s.v. 'crucifixion.'

Note: The BOTA Tarot keys have been reprinted with permission of the Builders of the Adytum, 5101 North Figueroa St., Los Angeles, CA 90042. (www.BOTA.org)

Note: The author has made every effort to give credit for images of the rosy cross used in this essay. The author and the editors cannot be held liable because they are being used for educational purposes.

Disclaimer: "Permission to use Builders of the Adytum images in no way constitutes endorsement of the material presented in this book."

Many thanks to Lisa Cucciniello for graciously editing this essay.

4

The Tradition of the Rose in East Indian Culture and Ayurvedic Medicine

By Monika Joshi

Roses have an enduring presence throughout East Indian cultural history. The flower was used and is still used today in ceremonies such as Hindu weddings, and in the practice of Ayurvedic medicine, a traditional system of Indian healing.[100] Applications of Ayurveda make extraordinary use of the blossoms, such as in preparation of rose fragrances. Often used in garlands, these floral decorations serve as a means of personal adornment. As a basic ingredient of perfumes, the incense industry makes use of the flower's derivatives. Rose petals are used in culinary preparations, such as those found in *The Great Curries of India*.[101] The purpose of this essay is to familiarize readers with the historical, religious and medical significance of roses in East Indian culture and Ayurvedic medicine.

Historical records about the varieties of native Indian

---

100 David Frawley, *Ayurvedic Healing: A Comprehensive Guide* (Delhi: Motilal Banarsidass Private Limited, 2000) 6.
101 For samples of recipes see *The Great Curries of India* by Camellia Panjabi, 40.

roses are very limited, except for ancient texts of Ayurveda, which identify the rose as a popular medicinal ingredient. However, within the last thirty years, archaeological discoveries indicate that flower was already in use in the early Indus Valley civilization as one of the basic ingredients for preparing aromatic oils. In 1975, Dr. Paolo Rovesti, director of the Instituto Derivati Vegetali in Milan, led an expedition to Pakistan, an independent country that was part of India until 1947, to investigate the early Indus Valley civilization. During his expedition, he found a perfectly preserved terracotta distillation apparatus in the Museum of Taxila, dating from about 3000 B.C. The presence of perfume containers exhibited in the museum, dating from the same period, confirms the preparation of aromatic oils in the Indus Valley civilization.[102] However, it was the fifteenth century large scale introduction and promotion that led to the popularity of roses throughout India. The credit for this promotion is usually given to the Mughal dynasty, a Muslim ruling family founded by Babur which ruled India from the early sixteenth century to the eighteenth century.[103]

Babur was a descendant of Genghis Khan, a thirteenth century ruler who founded an empire that included parts of China, Central Asia, the Middle East and Europe. In 1525, Babur established himself as sultan by marching his army from Kabul, Afghanistan, to Punjab, a region of the

---

102 For a more complete explanation of Dr. Rovesti's expedition, refer to *The Illustrated Encyclopedia of Essential Oils* by Julia Lawless, 18.
103 See *The Columbia Electronic Encyclopedia*, Sixth Edition, Columbia University Press, 2003, s.v. 'Babur.'

northwestern Indian subcontinent, and conquering all of India. Following Babur, the Mughal Empire produced five other great emperors, Humayun, Akbar, Jahangir, Shahjahan and Aurangjeb, who were collectively responsible for creating a massive empire in India. It was during the Mughal period that India saw a golden age of arts, military advancement, architecture and gardens.[104] Emperor Akbar, and later his son Jahangir, are credited with developing the Mughal style of art in India. Portraiture became prevalent in the time of Akbar, and he immortalized the rose through a portrait of himself holding the flower in his hand. Mughal art developed further during the time of Jahangir and his fascination with natural science resulted in paintings of birds, flowers and animals.[105] One of the most famous literary and art works that features a rose is "Gulistan, Garden of Sadi," a poetic manuscript commissioned by Shahjahan that contains paintings of roses Subsequent to the Mughals, the East India Company is credited with the introduction of a wide variety of roses into India. The East India Company was a commercial trading venture founded by the British in the 1700s. Eventually it would rule India as it acquired governmental and military functions. Roy E. Sheperd, a historian and author of "History of the Rose," an article for *Science Education*, writes that some species of roses have been cultivated by the East India Company in the botanical gardens of Calcutta.[106] Today, roses grace many of the gardens found throughout India.

---

104 Sarina Singh, *India*, 10th ed. (Oakland, CA: Lonely Planet, 2003), 21-26.
105 Ibid. 53.
106 Roy E. Shepard, "History of the Rose," *Science Education*, 39 (1954), 182-183.

Featured prominently in Shalimar Garden of Srinagar, the capital of Kashmir in the northernmost tip of India, a garden of roses is a legacy of Mughal emperor Shahjahan. This site demonstrates beautiful site contours through the channeling of a stream of water that flows down to the northeast corner of the lake from the hills above. Originally, the garden could only be accessed by water, but it is now accessible by a road that runs around the Dal Lake.[107] Coming south from Kashmir, there is *Zakir Gulab Bagh*, a famous rose garden in the area of Chandigarh. Dr. M.S. Randhawa, the first chief minister of Chandigarh and a patron of horticulture, is the mastermind planner of this plot. Established in 1967, the thirty acre garden is named after India's former president Dr. Zakir Hussain, and includes interplay of medicinal trees, fountains, and array of roses in different colors, shapes and sizes. This garden, with its ornately laid-out lawns and flower beds, has 1600 different species of roses which include both natural and hybrid varieties. Thousands of people visit its annual rose festival in February and March.[108]

Further south in Delhi, the capital of India, are the skillfully planned Mughal Gardens surrounding *Rashtrapati Bhavan,* the official presidential residence which are open to the public February and March of every year. The gardens are laid out on 130 acres and have 250 different varieties of

---

107 Attilio Petruccioli, "Shalimar Gardens in Srinagar," available from http://arch-net.org/library/sites/one-site.tcl?site_id=8861; Internet; accessed 19 March 2007.
108 Suni Systems (P) Ltd, "Chandigarh," available from http://www.webindia123.com/city/chandigarh/park.htm; Internet; accessed 19 March 2007.

roses. The Circular Garden area has 128 varieties of roses such as 'Tajmahal,' and 'Arjun.' The 'Oklahoma' variety has been described by a frequent visitor to this rose garden, as 'the darkest maroon rose', nearly black.[109] The landscape also features blue roses called 'Paradise' and 'Blue Moon,' as well as a rare variety of green roses.

Bangalore, a city in south India, serves as an industrial and educational center for the country. It is also known as the 'Garden City of India,' in honor of its beautiful gardens. Indian Muslim ruler, Hyder Ali, and later his son Tipu Sultan, commissioned a large botanical garden in the eighteenth century. Ali decided to create this garden, called 'Lalbagh,' as a means of rivaling the Mughal Gardens that were gaining popularity at that time. *Lal,* in Hindi, the national language of India, stands for red, and *Bagh,* stands for garden. It is said that Muslim emperor, Tipu Sultan, exclaimed, *"Lal Bagh!"* when he saw the garden's profusion of red roses along with its aesthetically designed lawns, flowerbeds, lotus pools and fountains. It is laid out on 240 acres of land and has hundreds of varieties of roses. It hosts annual flower shows which draw thousands of visitors.[110] Along with being a means of visual pleasure in the botanical gardens previously mentioned, roses are used in their natural form in Indian the elaborate wedding ceremonies that are popular throughout the country.

---

109 Eyewitness account from Raj Kumari Sharma, the author's mother, 14 February 2007.
110 Lonely Planet Publications, "Lalbagh Botanical Gardens," available from http://www.lonelyplanet.com/worldguide/destinations/asia/india/bangalore?poi=1000151254; Internet; accessed 14 April 2007. Singh, *India,* 53.

These lavish occasions provide a great market for fresh flowers, especially roses. The use of the flowers begins with the decoration of the floral archway that leads to the wedding hall. These arches are covered with flower petals that are arranged to form beautiful patterns, such as the names of the bride and groom. During the ceremony, the bride and groom often exchange rose garlands, and the wedding *mandap*, a specially prepared stage or enclosure, is often decorated with roses. This structure contains support beams from which strings of flowers and mango leaves are hung to form an elaborate canopy where the wedding ceremony takes place. After the wedding, a grand reception takes place where the entire background of the stage for the bride and groom is decorated with countless strings of flowers. The wedding night also makes abundant use of the roses in decorating the nuptial bed. Canopied beds are specially prepared with flowers; the most preferred being roses, in order to create a curtain from all four posts. Rose petals are strewn on the marital bed to create an inviting and memorable setting for the young couple on their first night together. Along with wedding rituals, roses are widely used in religious practices.

In worship ceremonies, garlands of roses are frequently offered to deities. Such an offering is the favorite of the Hindu God Krishna. In his book *Sacred Waters,* western author Stephen Alter recounts personal experiences of using roses in worship. On a pilgrimage to the Ganges River, the most sacred river in India, a priest gave him a handful of flower petals, jasmine, rose and marigold, in his cupped palms and told him to release them into the Holy Ganges. He also recounts that he was instructed to sprinkle rose petals on the statue of Krishna.[111] During a visit to *Vrindavan*, 135 kilometers south of Delhi in the state of Uttar Pradesh, a place renowned for its temples of Lord Krishna, an eyewitness saw the roads leading to the temples filled with vendors on both sides. The rose vendors were ready with beautiful, fragrant roses before devotees started pouring in for morning prayers. The scent of the roses was so intoxicating, that it stayed with the visitor on the long drive back to Delhi.[112]

According to the Brahma Kumari World Spiritual organization, which originated in India, roses are symbolic of surrender. The falling of each rose petal, one by one, is compared to the human surrender to God in all aspects of life. Shree Kripaluji Maharaj, founder of *Jagadguru Kripalu Parishat* and spiritual master and a poet-saint of recent times, has influenced thousands of devotees around the world in the *Bhakti Marg*, the path of love or devotion to God.[113] He calls the rose, "King of Flowers," for its beauty and fragrance. Along with religious ceremonies, the rose also has a place in Indian folklore.

111 This river is sacred to Hindus and according to popular belief, to bathe in the sacred waters of the river is to wash away an individual's sins. Another belief states that scattering the ashes of the dead into the river will ensure that the soul reaches heaven. Several temples have been built at strategic points along the river in order to attract large numbers of pilgrims year-round.
112 Eyewitness account by the author on a visit to India 28 March 2004.
113 Hindu philosophy describes four paths to God realization. These paths are Bhakti yoga, Karma yoga, Raja yoga and Hatha yoga. Bhakti yoga or marg (path) is the path of love or devotion to God. Karma yoga is the path of service. Raja yoga or royal (path) is the path of scientific meditation and Hatha yoga is the path of disciplining the body to transcend the body's consciousness.

While the lotus finds mention in the early Vedic texts of ancient India, roses are rarely featured, with the exception of the legend of *Padmavati,* dated much later. According to an ancient Hindu legend, the goddess *Lakshmi* is said to have been created from 108 large and 1008 small rose petals. The story relates that both God *Brahma* and God *Vishnu,* had an argument as to which was more beautiful, the lotus or the rose?[114] While Vishnu said it was the rose, Brahma, who had never seen a rose, refused to accept it. Finally, when Brahma saw a rose, he was so enamored with the beauty of the flower that he created the goddess Lakshmi out its petals.[115] As mentioned earlier, the traditional form of Ayurvedic Indian healing is steeped with the tradition of the rose.

The language of Ayurveda is Sanskrit, one of the oldest written tongues. The roots of the word are *Ayus* and *Vedas.* *Ayus* may be translated as 'life' and *Vedas* as 'knowledge.' Hence, Ayurveda is the 'Knowledge of Life.'[116] According to astronomical records in the ancient Vedas, the most authoritative Hindu sacred texts, Ayurveda was in practice before 4000 B.C. Ayurveda teaches that man is a microcosm, a universe within himself. He is a child of the cosmic forces of the external environment, the macrocosm. His individual existence is indivisible from the total cosmic manifestation.[117]

114 The Hindu trinity consists of Brahma, the creator; Vishnu, the preserver, and Shiva, the destroyer.
115 Rosefarm.com, "A Brief History of Roses," available from http://www.rose-farm.com/history.php; Internet; accessed 15 March 2007.
116 Marc Halpern, *Principles of Ayurvedic Medicine, Student's Textbook, Part One,* 6th ed. (Grass Valley, CA: California College of Ayurveda, 2004) 1-3.
117 Frawley, *Ayurvedic Healing: A Comprehensive Guide,* 6.

Ayurveda views health and "dis-ease" in holistic terms, taking into consideration the inherent relationship between the individual and the cosmic spirit, individual and cosmic consciousness, and energy and matter.[118] The wisdom of Ayurveda was intuitively received in the hearts of *rishis,* the seers of truth. They perceived that consciousness was energy manifested into the five basic principles, or elements; Ether or space, Air, Water, Fire and Earth. This concept of the five elements is the heart of Ayurvedic science, which states that the five basic elements present in all matter also exist within each individual and manifest themselves in the functioning of the five senses of hearing, touch, vision, taste and smell. The physiology of the human being is governed by three main forces which are commonly known as the three *doshas* or humors, called *Vata, Pitta* and *Kapha. Vata* is made up of air and ether, *Pitta* is made up of fire and water and *Kapha* is made up of water and earth. These three humors, govern all the biological, psychological and physiological functions of the body, mind and consciousness. If they are out of balance, the result is *disease.*[119]

Ayurveda recognizes the unique physical and mental constitution of an individual and detects the current state of imbalance through various techniques. It then offers a wide range of modalities to enable individuals to take care of their health. Ayurveda is not a system of healing in which everyone

---

118 Vasant Lad, "Ayurveda, The Science of Self Healing" (Delhi: Motilal Banarsi-dass Publishers, 1984), 18.
119 Frawley. Ibid.

enjoys the same practices and rarely are two treatment plans alike. Healing through Ayurveda involves all five senses, and the rose is used in many of these treatments. The sense of taste is addressed through proper diet and herbs, and roses are often combined with herbs which are then administered as an integral part of the healing process. Aromatherapy addresses the sense of smell, of which rose fragrances are often present. Vision is addressed through color therapy, and roses can be utilized for this purpose. Music and sound energies apply to the sense of hearing. Finally, the sense of touch is administered to through specially prepared herbal oil massages, of which rose oil is often an ingredient.

The cooling and soothing qualities of roses are used in various treatments. Rose petals, with their sweet taste and cool properties, have an excellent post-digestive effect. Roses are also used in laxative and digestive formulas because they enhance absorption. Emotionally, roses help increase feelings of love, compassion and devotion. Rose derivatives, like rose petal powder, massage oil, or aromatherapy oil, are alterative, antimicrobial and antipyretic. They cleanse the eyes and nourish the female reproductive system.[120] Cold poultices with rose and sandalwood pastes are sometimes prescribed to alleviate any kind of burning sensations. When used over time, the "essence of rose" pacifies heat in the blood, thus keeping imbalances in the skin from erupting. This is often used in acne treatments in order to remove any dullness in

---

120 Halpern. Ibid. 10-17.

the skin. Rose derivatives are unsurpassed beauty oils. Rose-based oils benefit every skin type and are often used in lotions to treat infected, dry, or sensitive skin. Conjunctivitis, eye pain, irritation and various inflammations of the eyes can be treated with the use of rose water.

Aromatherapy is an integral part of Ayurveda. Unlike foods or herbs, aromas work more directly on the mind and body and are often an important part of treating psychological disorders.[121] According to Ayurvedic texts, roses are connected with physiological factors that govern emotions and their effects on the heart. Rose extracts are soothing to both the heart and its emotions. The fragrance of rose oil produces a gentle but effective antidepressant effect. It brings joy to the heart, promotes feelings of love, reduces fear, drives away melancholy, helps recovery from grief and also settles the heart by combating the unwillingness to let go.[122] As a perfume, rose fragrances represent the essence of purity and innocence, yet can also act as an aphrodisiac. *Gulkand,* an Ayurvedic jam made primarily of rose petals and rejuvenating herbs, is used as a cooling tonic to combat fatigue and heat-related conditions; it is also naturally rich in calcium and has antioxidant properties.[123]

The ancient science of Ayurveda and its rose-focused prescriptions continue to be used effectively in modern days and can be applied anywhere in the world. One example

121 Ibid. 10-12.
122 Frawley. Ibid. 327.
123 David Crow, "The Pharmacy of Flowers," available from http://www.floraco-peia.com/article.php?article=10; Internet; accessed 17 March 2007.

occurred in Fresno, California, 18 January 2007. A terminally ill eighty-seven year old hospital patient had liver cancer and was given two weeks to live. The family requested an Ayurvedic practitioner to help the man through his transition from this life. Rose water was sprayed around the patient and soothing chants were recited in the background. This rose therapy helped the man to be calm and therefore well rested, and even helped his roommate sleep through the night for the first time. The nurses were amazed, but it is believed that the high frequency effects of the rose water and chants created this transformation of atmosphere in the room and ultimately in the patients.[124]

Indian culture has a strong emphasis on religion and spirituality and gives much importance to the role that aromas can play in evoking divine inspiration. It is believed that the aroma of sandalwood, jasmine or rose can carry the human spirit to sublime levels. Rich aromas of plants, herbs and flowers have been tapped and incorporated as pure essences, often in oils, to create an atmosphere conducive to spirituality. In addition, these fragrances, in the form of anointing liquids, whether oil or water based, are used in personal adornment. Perhaps one of the oldest texts that record the use of fragrant oils for this purpose is the *Kamasutra*, the comprehensive text on sex and sexuality attributed to the sage *Vatsyayana* in the early fourth century A.D. The use of perfumed oils is listed as one of the sixty-four arts of which women should possess

124 Eyewitness account by the author, 18 January 2007.

great knowledge. *Vatsyayana* also counsels that an ideal man would apply a limited quantity of ointments and perfumes to his body.[125] In another chapter, the sage talks about the use of fresh flowers and ointments as a means to create the atmosphere for sexual congress. Reference to "the pleasure-room, decorated with flowers, and fragrant with perfumes" appears in the voluminous treatise on sexuality.[126]

An Indian legend attributes the discovery of rose oils to the Mughal emperor Jahangir in 1612. According to the legend, the palace gardens of the Emperor Jahangir were interspersed with canals and fountains. Rose petals were strewn on the water to form a picturesque spectacle for special occasions. The pools and ponds of his estate were decorated with rose petals in preparation for his wedding at which point a film of oil was seen floating on the water. One version of the legend says that it was Jahangir who noticed the oily film, and another version credits his betrothed. The royal personages were so enamored with the scent of the oil that they ordered that the oil be bottled for later use; thus began the use of rose oil perfume. However the discovery of the ancient distillation apparatus mentioned earlier proves that rose oil was used much earlier than this seventeenth century legend.

---

125 Richard Burton, "On the Arts and Sciences to be Studied," in *Kamasutra* [book on-line], available from http://www.kamasutra-sex.org/text/kama103.htm; Internet; accessed 17 March 2007.
126 Richard Burton, "On How to Begin and How to End Congress" in *Kamasutra* [book on-line], available from http://www.kamasutra-sex.org/text/kama210.htm; Internet; accessed 17 March 2007.

Today, production of high quality rose oil is labor intensive. Compared to other plants such as lavender and rosemary, rose petals secrete an infinitesimal amount of essential oil. It takes 3,500 kilos of flowers to produce one kilo of oil. This amounts to 1,400,000 handpicked blossoms to produce thirty five kilos of oil, 40,000 blossoms for one ounce of oil, and sixty-seven blossoms to make one drop of oil. Harvests must be done in the early morning hours before the heat of the sun evaporates the oil.[127] Roses, like all flowers, have a biorhythm that dictates cycles of blossoming and the production of fragrance molecules. The harvesting of roses for distillation begins in the early dawn at a time the Vedas call *Brahma Muhurta*, 'God's time.' At these pre-dawn hours the rose petals are at the peak production of their oils. Roses produce their maximum levels of damascenone, the primary molecule of rose fragrance, on the mornings of a full moon.[128] According to David Crow, author of *Floracopia*, more than 5,000 varieties of roses exist, yet only a few give the fragrance sought by perfumeries, the most popular being pink to light red 'Damascus' roses. A rose specialty 'Attar of Rose' is a traditional perfume of India that is composed of essential oils distilled into a base of sandalwood oil. The process of distilling roses for attar is rather simple; roses are harvested in April and May and placed in large copper vessels which are connected to the receiving vessel with a bamboo pipe that is covered with special grass rope for insulation. Flowers are

127 Crow. Ibid.
128 Ibid.

placed in water-filled distilling vessels and are then heated. The vapors pass to a receiving vessel containing sandalwood oil, which absorbs the rose essence over a period of fifteen to twenty days, with new flowers being distilled every day. Rose attar, with its rich fragrant and intoxicating aroma, was often used as an aphrodisiac in ancient perfume aromatics and anointing oils.[129]

Today there is a concerted effort by the Indian government to step up the production of roses, particularly for the fragrance industry. It is being promoted heavily by government research institutes, such as the Institute of Himalayan Bio Resource Technology, which is engaged not only in the cultivation but also in the development of better and more affordable distilling processes.[130] The prominence given to this flower by leaders and celebrities has ensured that the rose continues to hold a special place in the country. The most famous personality of India, the first prime minister, Pandit Jawaharlal Nehru, always sported a rose in his buttonhole.[131] Movie personalities, politicians and other celebrities are often honored with garlands of roses from fans striving to outdo each other in the quantity of the flowers. A famous South Indian movie actor, Rajanikanth, made

---

129 See *The Columbia Electronic Encyclopedia*, Sixth Edition, Columbia University Press, 2003, s.v. 'Attar of Roses.'

130 Christopher McMahon, "More about Sacred Oils and Pure Essential Attar Oils © Exclusively at Erzulie's" available from http://www.erzulies.co.uk/site/articles/view/22; Internet; accessed 17 March 2007.

131 Thiru, Dayanidhi Maran, "Speech of Thiru. Dayanidhi Maran Hon'ble Union Minister of Communications and Information Technology at The Release of Stamps on 'FRAGRANCE OF ROSES'" at Hotel Taj Lands, Mumbai, held on 02-07-2007" available from http://www.dmaran.nic.in/speechdisplay.php?id=207; Internet; accessed 18 February 2007.

headlines on his forty-ninth birthday in December of 1998, when he was presented with a garland of red roses weighing 25 kilograms.

A view of the culture of India would not be complete without mention of the Indian movie songs that make ample use of roses in their lyrics. *Gulaabi aankhen jo teri dekhi sharaabi yeh dil ho gaya,* are lyrics of a song sung by famed playback singer Mohammed Rafi in the movie *The Train.* The phrase translates to, 'Looking at your beautiful rose-like eyes, my heart is drunk with your beauty.'[132] Another song from the film *Mantra,* sung by Faakir is... *Aankhon main shabab hai, chehra yeh gulab hai...* which means, 'Your eyes have a magic and your face is like a rose.' There is an interesting blend of diversity, rich culture, religion and tradition that permeates every aspect of life in India – with the rose being a significant part. In fact, India has given a lasting tribute to the beauty and fragrance of roses by unveiling scented stamps titled *Fragrance of Roses* to commemorate Valentine's Day in February 2007. One must consider the following scenarios of daily life in India.

In Delhi, the capital city of India, crowded bazaars of small shops are placed side by side flanking the narrow roads leading to the temples. Baskets of colorful roses vie with garlands of jasmines and marigolds. Women from all areas sport jasmines, marigolds or roses in their well-oiled,

132 Hindilyrix.com, "Gulaabi Aankhen," available from http://www.hindilyrix. com/songs/get_song_Gulaabi%20Aankhen.html; Internet; accessed 14 March 2007.

long, black hair. In the center of the most holy inner court of the temple is the statue of a deity covered in garlands of flowers. Priests chant sacred hymns while showering petals onto the statue of the deity. The choice flower is the rose, premier among all other floral species. After the worship ceremony, the priest distributes sacred water, vermillion powder, and petals or whole flowers to all in attendance. Women graciously accept the flowers, vying for the roses which they carefully tuck into their hair on their way out of the temple. In one more scenario, crowds gather around a priest on the banks of River Ganges. The priest solemnly chants sacred *mantras*; sacred words which are said to have originated from the sacred Hindu text, the Vedas.[133] Hindus believe that these words have powers when uttered with the right intonation and bring many blessings to them when used as part of any ritual. The priests also have specific gestures that accompany the assemblage to strew the petals they are holding in their cupped hands into the river. Upon being release, the red rose petals float gently on the surface of the river as they are carried away by the current.

Roses play an important role in Indian culture and will continue adorning buttonholes, peeping out from black tresses, and adding their own color to the vibrant sub-continent. But more than just being a personal ornament, the rose is valued for its healing properties, such as in Ayurvedic

---

133 See *Britannica Concise Encyclopedia.* Encyclopedia Britannica, Inc., 2006, s.v. 'mantras.'

treatments, and religious traditions, such as adorning statues of Krishna. The rose, with all of its properties, permeates the daily culture of East Indian life.

5

Rose to Rosary: The Origins of the Flower of Venus in Catholicism

By Lisa Cucciniello

Centuries of Roman Catholics have shown their piety and dedication to the Virgin Mary by praying the rosary. Pope Leo XIII, who reigned from 1878-1903, referred to the rosary as "this pious method of prayer." Several decades later, one of Leo XIII's successors, Pope Pius XII, who reigned from 1939-1958 explained, "Such piety towards the blessed Virgin is the hallmark of a profoundly Catholic heart."[134] Throughout Catholic history, endorsement of the rosary by church leaders helped make this devotion an enduring Catholic icon. But the *origin* of the Catholic rosary and its connection to the actual flower, the rose, remain obscure. In fact, the rosary evolved over centuries, beginning in 753 BC with the founding of Rome and continuing to the present time when the late Pope John Paul II added his touch to this

---

134 Pope Paul VI, Pope John XXIII, Pope Leo XIII. <u>Seventeen Papal Documents on the Rosary</u>. (NY: Daughters of St. Paul, 1980) 111. *The Holy Rosary: Selected and Arranged by the Benedictine Monks of Solemes*. Trans. Rev. Paul J. Oligny, OFM (St. Paul Editions: Daughters of St. Paul, 1980), 230.

important symbol in 2002, causing some contention among Catholic congregations.

What began as a simple rose serving as a symbol of the pagan Roman goddess Venus ultimately transformed into a Catholic icon glorifying the Virgin Mary. Part of this metamorphosis included using both the rosary and the Virgin Mary as a theme in Medieval and Renaissance artwork. Publicity of the rosary would help promote the Catholic faith and maintain the dedication of the congregation, especially when the Catholic Church was in danger of losing members, such as during the Albigensian Crusade, the ravaging of Europe by the Black Plague and the Protestant Reformation. Active sponsorship of the rosary can be credited to the Dominican order of Catholic priests, an order that historically traces its roots to St. Dominic's preaching in the early twelfth century in the Languedoc area of Southern France.[135] The Dominican endorsement of the rosary is described throughout this chapter, with prominent Dominican leaders of the Catholic Church actively involved with the sponsorship of the rosary. In the fifteenth century, Dominican preacher Alan de la Roche vigorously promoted

---

135 See *Catholic Encyclopedia*, Volume XII, 1911, s.v. 'Dominicans.' The *Catholic Encyclopedia* is a compilation of over 10,000 articles which, as its preface states, "proposes to give its readers full and authoritative information on the entire cycle of Catholic interests, action and doctrine." It does not limit itself to Church doctrine, but also to significant contributions such as art and charity work made by Catholics throughout history. Dominicans are one of the many orders of priests and nuns in the Catholic Church. For a more comprehensive list of Catholic religious orders, see *A Guide to Religious Ministries for Catholic Men and Women*, an annual publication by the Catholic News Publishing Company, which enumerates with a brief overview of the history of the many orders of priests and nuns in the Catholic faith.

the rosary throughout France and the Netherlands, deeming it his "special mission."[136] Contemporary to Roche was Father Jacob Sprenger, a Dominican priest who founded the first Confraternity of the Holy Rosary in 1475 in order to actively sanction commitment to the rosary, and Catholicism overall.[137] Along with a chronology of the Dominican approval of the rosary, this chapter also offers a possible explanation as to why the *rose* was chosen as an emblem to honor the Virgin Mary, even though this symbol is not unique to Christianity. Surprisingly, the rose as an icon of veneration has roots in the pre-Christian Era, and was originally used by the Ancient Romans as a symbol of devotion to the Ancient Roman goddess Venus.

The founding of the City of Rome in 753 BC marked the beginning of the Roman Empire, which would become one of the great civilizations of history. Until Emperor Constantine the Great Christianized Rome in the fourth century AD, Rome was pagan; its citizens worshipped more than one god, making them polytheistic.[138] An important symbol of Roman pagan rituals was the rose. The rose was the flower of Venus, an ancient Roman goddess derived from the earlier Greek goddess Aphrodite. Initially, Venus was the goddess of

---

136 See *Catholic Encyclopedia*, Volume I, *1907*, s.v. 'Alan de la Roche.'
137 See *Catholic Encyclopedia*, Volume III, 1912, s.v. 'Confraternity of the Holy Rosary.'
138 The term 'Constantine the Great' is used by the *Catholic Encyclopedia*, Volume IV, published in 1908. In 325 AD, Constantine called the Council of Nicea, which helped unify Catholic doctrine, discussing issues such as the marital status of priests, excommunication, and the core beliefs of Catholicism. According to the *Catholic Encyclopedia* Volume XI, published in 1911, paganism is any religion outside of Christianity, Judaism and Islam.

cultivated fields and gardens, but would later be renowned as the goddess of love and beauty, and furthermore revered for her wisdom.[139] As a result of Constantine Christianizing Rome, the Catholic Church became more prominent throughout Europe, and the rose would come to honor the Virgin Mary and her role in the life of Christ, thereby veiling its pagan roots. It is at this time that Europe begins to see the adjustment of former pagan beliefs to fit a new Christian identity. The earlier pagan Roman rose would now being associated with the Virgin Mary as a result of the Christianization of Rome. However, according to John S. Stokes Jr., a Roman Catholic member of the Society of Mary, "The association of plants and flowers with the Blessed Virgin Mary originated with the early Church Fathers, who saw her prefigured in passages from the Old Testament containing nature imagery."[140] In the *Bible*, Canticles 2:1 states, "I am a rose of Sharon, a lily of the valley." Stokes further explains that "from this period, also, comes the legend that after Mary's Assumption into heaven, roses and lilies were found in her tomb."[141] As recent as 1955, Pope Pius XII addressed a group of rose growers in Rome and discussed the pagan origins of the rose and its transformation into a symbol of Christianity stating, "When the memories of paganism were erased, the charm of the rose reverted to the

139 *The Columbia Electronic Encyclopedia*, Sixth Edition. NY: Columbia UP, 2003.
140 John S. Stokes, "Flowers of the Virgin Mary," AVE, *Society of Mary*, (London, 1984) 1.
141 Ibid.

true God."[142] The rose as a symbol of reverence underwent an evolutionary process that would transform it into the *rosary*. As the Catholic Church's power was threatened by heresy, disease, and, later, the Protestant Reformation, the rosary would be more aggressively promoted starting with the Albigensian Crusade in 1198.

For many years Catholic tradition traced the origin of the rosary to St. Dominic, founder of the Order of the Preachers or Dominicans devoted to ministering to and saving souls. Tradition acknowledged that in 1198, St. Dominic defeated the Albigensians, an heretical group that prevailed in southern France during the twelfth century whose teachings were contrary to those of the Catholic Church. According to the Albigensians, life was a dichotomy of good and evil, where good created the spiritual world and evil created the material world; thus, man was a living contradiction since he lived in the material world while seeking salvation in the spiritual world. Because of this conundrum, the Albigensians concluded that liberation of the soul was the true end of being, and the most commendable death was suicide by starvation. In contrast, the Catholic Church taught, and still teaches that only God has "direct dominion over life," therefore it is a sin for one to end his time on Earth prematurely, such as by committing suicide.[143] According to Pope Innocent III, who launched the Crusade against the Albigensians, the Albigensian heretics

142 Pope Pius XII, "To an International Group of Rose Growers in Rome," *The Pope Speaks* 2 (Summer 1955) 133-135.
143 See *Catholic Encyclopedia*, Volume XIV, 1912, s.v. 'Albigensians.'

were more reprehensive than the Saracens, the Muslims whom the Catholic Church battled throughout the majority of the Crusades. Though the Crusades were fought primarily against the Muslims in order to recapture Jerusalem, the Catholic Church did wage war against the Christian Albigensians. The Catholic Church saw the Albigensian heresy as a force that would destroy the entire human race; therefore, the group had to be abolished and a Crusade was launched to put an end to the heretical teachings. Catholic tradition taught that during the Albigensian Crusade, the Virgin Mary interceded on behalf of the Catholics as a result of St. Dominic praying the rosary, causing the Catholics to emerge victorious over the Albigensians.[144] Leaders of the church continually sanctioned this legend throughout history:

Pierre Foldes, volunteer with Doctors of the World, October 1999.

The well-known origin of the Rosary, illustrated in the celebrated monuments of which we have made frequent mention, bears witness to its remarkable efficacy. For in the days when the Albigensian sect, posing as the champion of pure faith and morals, but in reality introducing the worst kind of anarchy and corruption, brought many a nation to its utter ruin, the Church fought against it and the other infamous factions associated with it, not with troops and arms but chiefly with the power of the most Holy Rosary, the devotion which the Mother of God taught to our Father Dominic, in order he might propagate by it. (Pope Leo XIII 1892)

---

144 The *Catholic Encyclopedia*, Volume I, 1907, s.v. 'Crusades.'

However, the tradition of St. Dominic's connection to the rosary did not emerge until the latter fifteenth century. The legend began with Alan de la Roche, a distinguished member of the Dominican Order who was awarded the high honor of Master of Sacred Theology in 1473, a professional degree granted by Dominican seminaries.

Roche circulated the legend of St. Dominic and the conquest of the Albigensians to advocate the reputation of the Dominican order over the Carthusian order, another Catholic priestly community that committed much of its time to private prayer. In contrast, the Dominicans believed in active preaching to the outside community. Guy C. Bauman, writing in the *Metropolitan Museum Journal,* argues that an already existing strain between the two orders incited Roche to make a claim that would distinguish the Dominicans from the Carthusians.[145] Today, the tradition of St. Dominic and his defeat of the Albigensians is accepted as folklore, and some scholars, such as W.A. Hinnensbusch of *The New Catholic Encyclopedia,* now agree that the rosary's connection to St. Dominic is "apocryphal."[146] Perhaps the folklore was fostered because the Carthusian order had already begun to influence the Catholic faith by making daily devotion outside the formal mass more accessible to the mass illiterate

145 Guy C. Bauman, "A Rosary Picture with a View of the Park of the Ducal Palace in Brussels, Possibly by Goswijn van der Weyden," *Metropolitan Museum Journal,* 24 (1989), 138. The *Catholic Encyclopedia,* Volume III briefly explains the Carthusian order as being a small, slow growing order of monks whose daily life greatly comprised of solitary meditation and reflection. In contrast, the Dominicans were an active preaching order dedicated to the salvation of souls.
146 W.A. Hinnensbusch, "The Rosary," *The New Catholic Encyclopedia,* Boston. *Seventeen Papal Documents on the Rosary,* Daughters of St. Paul, 667.

population. It could easily be understood then why Roche
would connect St. Dominic, the founder of the Dominican
order of priests, to such an important event in Catholic
history as the Albigensian Crusade.

Anne Winston, Associate Professor of German at
Southern Illinois University at Carbondale, explains that
Dominic of Prussia, a Carthusian monk from Poland
noted for his intelligence and religious fervor in the early
fifteenth century, predates Roche's claim that the origins of
the rosary rest with St. Dominic. In his own work, *Liber
Experimentiarum*, Dominic of Prussia tells how he reflected
on a series of meditations on the life of Jesus and the Virgin
Mary while praying a succession of 'Hail Marys.'[147] During
the time of Dominic of Prussia, the 150 Psalms of the *Bible*
were regularly recited outside the formal mass as a form
of devotional prayer. However, since the majority of the
population of the time was illiterate, many were excluded
from this practice, as memorizing 150 individual Psalms
without the aid of a prayer book was exceedingly complex.
Consequently, some Catholics began substituting the 150
Psalms with repetitive 'Hail Marys' or 'Our Fathers,' since
these prayers were known by most Catholics and easier to
commit to memory. This alternate form of recitation became
known as the 'Psalter of Mary.'[148] According to the *New
Catholic Encyclopedia*, Dominic of Prussia popularized

---

147 Anne Winston, "Tracing the Origins of the Rosary: German Vernacular Texts,"
   Speculum Vol. 68 (Jul. 1993) 622.
148 See *Catholic Encyclopedia*, s.v. 'rosary.'

this method of affection to the Virgin Mary as early as 1409, substituting the fifty Psalms of the *Bible* that spoke specifically of Jesus and the Virgin Mary with fifty 'Hail Marys.' This practice of saying fifty consecutive 'Hail Marys' became known as a *rosarium*, the Latin term for rose garden, which was often used during the fifteenth century to describe a succession of repetitive prayers.[149] By the time of Dominic of Prussia and Alan de la Roche in the later fifteenth century, an association between the rose itself and the Virgin Mary had already been established by the early church leaders, but further circulation of the *rosary* would help veil the earlier pagan connotations of the *rose*. Part of this process would be the transformation of the *object* used to show dedication. Pagan Romans used an actual *rose*, but Catholics grew to use mostly *rosary beads* for their devotional. However, the namesake of the Catholic rosary, the rose, remains.

A main component of the rosary dedication is the string of beads held in one's hands while the repetitive 'Hail Marys,' 'Our Fathers,' and 'Glory Bes' that make up the standard prayers of the Catholic rosary are recited. At one time, the individual beads were in the shape of little roses, the flower for which the dedication is named.[150] Each single bead represents a specific prayer of the rosary, and the prayers are said in a

---

149 *New Catholic Encyclopedia* Volume XII (NY: McGraw Hill Book Company, 1967) 669.

150 Today the rosary is most commonly made of oval or round beads, most likely for economic purposes. While the beads are a single string, terms such as "a pair of rosary beads," "a set of beads," or "a pair of beads" are sometimes used. The origins of the phrases are uncertain, but perhaps they refer to the fact that the rosary beads branch out from a common point, a crucifix, then split into two and connect to form a circle.

particular order according to the proscribed tradition of the devotion. As worshippers progress through the prayers, they pass one bead through their fingers for each prayer said, in order to keep count of prayers already recited. Like the rose, the tradition of using objects to keep track of prayers is not uniquely Catholic. Though *rosary* beads are often associated with Western Europe and Catholicism, *prayer* beads date back to the third century in areas of the East, where they were used in other faiths such as Buddhism and Islam.

During the third century, Paul of Thebes, who is recognized as the first Catholic hermit, used stones to keep track of his daily ritual of saying three hundred 'Our Fathers.' According to Catholic tradition, Paul was born in Egypt and fled into the wilderness during the persecution of Christians by the Roman Emperors Decius and Valeranius circa 250 AD. He resided in a cave for ninety-one years, where he devoutly practiced his faith until he died at the age of 114.[151] Saint Anthony, a wealthy man of the third century, used pebbles and knotted strings to keep track of his daily recitations when he was in the deserts of Egypt and Syria. When St. Anthony heard the words from the Gospel of Matthew, "If thou wilt be perfect, go and sell all thou hast," he felt they were directed to him and he renounced his worldly belongings to lead an austere life in the desert.[152] Marco Polo, a native of Venice who traveled to present day China in

---

151 Bauman, "A Rosary Picture with a View of the Park of the Ducal Palace in Brussels, Possibly by Goswijn van der Weyden."
152 See *Catholic Encyclopedia*, s.v. St. Anthony.

the thirteenth century, wrote of the Buddhist King Malabar wearing a chain of 104 gems, which the king used to keep track of his daily devotions. Islamic tradition used strings of ninety-nine prayer beads to represent the ninety-nine names of *Allah,* a tradition that continues in the present day. This Muslim ritual appears to have its roots in the ancient Hindu god Shiva, who is worshipped as both the creator and the destroyer.[153] Similar to these earlier traditions, Catholics used prayer beads to help keep track of recited prayers. However, Catholics eventually chose the term *rosary* beads instead of *prayer* beads, a reflection of their continual embrace of the primeval imagery of the rose. In addition to the name of the devotion that distinguishes this Catholic ritual from other traditions, the Catholic rosary consists of a series of fifteen meditations called mysteries, which are reflected upon as the devotee progresses through the prayers of the rosary.[154] These meditations focus on important events in the life of Jesus Christ and the Virgin Mary, and are taken from the Gospels of Matthew and Luke and the Acts of the Apostles. The reflection on these mysteries is fundamental to the Catholic rosary; many of the core beliefs of the Catholic faith, such as the death and resurrection of Jesus, are reflected upon throughout the recitation of the rosary. As people recite the

---

153 *Encyclopedia Britannica* 2006 explains that the word *Allah* is the standard Arabic word meaning God. Some of the ninety-nine names include, "the One and Only," "the Living One," and "the Real Truth." Anne Winston-Allen. *Stories of the Rose: The Making of the Rosary in the Middle Ages* (University Park: The University of Pennsylvania Press, 1997), 14.

154 In 2002, Pope John Paul II sanctioned five more events in the life of Christ which he called the Luminous Mysteries. For the purpose of this essay, mainly the Joyful, Sorrowful, and Glorious Mysteries will be discussed.

standard consecutive prayers of the rosary, they are not just repeating prayer after prayer but also reflecting on the events of the mysteries and what their significance is to their faith.

As one prays the rosary and reflects on the meditations, he or she is taken through the crux of Catholic beliefs. Rev. Paul J. Oligny, who translated *The Holy Rosary: Selected and Arranged by the Benedictine Monks of Solesmes*, explains in the forward, "Truly no more urgent and reliable education and re-education in the faith can be conceived than the rosary, the whole rosary that has become a habit. The Creed becomes what it truly is: a prayer. There we find the faith, whole and living, the true faith, the faith that begins with humility and ends in praise."[155] The Creed to which Oligny refers is the Apostle's Creed, a prayer believed to be composed by the original twelve apostles of Jesus forty days after His [sic][156] death. Catholic belief states that on the fortieth day after Jesus rose from the dead, the Holy Spirit descended upon the twelve apostles and inspired them to publicly preach His mission. Holy Spirit revealed Himself as tongues of fire over the apostles' heads, and at this time they composed the Creed. In doing so, each apostle contributed one of the articles, or beliefs, that are central to Catholicism.

---

155 *The Holy Rosary: Selected and Arranged by the Benedictine Monks of Solesmes*, trans. Rev. Paul J. Oligny, OFM, 14. Oligny states that an entire education of the faith can be learned from the devotion of the rosary because of the prayers, meditations, and the beliefs expressed in the Creed. See *Catholic Encyclopedia* Volume I for a detailed explanation of the Apostle's Creed.

156 In Catholic tradition, when referring to God or Jesus, using words such as Him, His, or He, the denotations are capitalized, as explained by the *Catholic Encyclopedia*, s.v. 'Jehovah,' because they refer to the sacred name of God or Jesus.

To begin the prayer of the rosary, the devotee takes the rosary beads in his or her right hand and makes the sign of the cross. After the sign of the cross is made, the Apostle's Creed is recited, followed by one 'Our Father,' three 'Hail Marys' and one 'Glory Be.' The first mystery is read, followed by an 'Our Father,' a decade, or progression of ten consecutive 'Hail Marys' and another 'Glory Be.' As worshippers proceed through each decade, they reflect on the mystery and its significance assigned to that particular decade. This pattern of prayers continues for each of the mysteries. After the recitation of an entire set of mysteries, the rosary ends with the 'Hail Holy Queen,' an honorary prayer praising the Virgin Mary as the Mother of God and asking for her mercy on sinners. While there are fifteen mysteries in total, it is not obligatory to reflect upon all fifteen daily.

The *Catholic Encyclopedia* explains that certain sets of mysteries are reflected upon on certain days of the week. On Mondays and Saturdays one reflects upon the Joyful Mysteries. This series consists of the Annunciation, where the Virgin Mary was told by the angel Gabriel that she was going to bear Jesus; the Visitation, where the Virgin Mary visits her cousin Elizabeth, who in her old age, is miraculously bearing Jesus' cousin later known as John the Baptist; the Nativity, which is the story of the birth of Jesus, the Presentation of Jesus at the Temple, where in accordance with Jewish law, Jesus was presented at the Temple after being born; and the discovery of Jesus at the Temple, where Jesus is found at the age of twelve by His parents, preaching beyond His years to the elders of the Temple.

In 2002, an unexpected change was made to the centuries old tradition of the rosary when Pope John Paul II added the Luminous Mysteries, a series of events taken from the four Gospels of Matthew, Mark, Luke and John that mark significant events in Jesus' public ministry. If one was to order the mysteries chronologically, according to when the events occurred in the life of Jesus, the Joyful Mysteries, which include Jesus' birth and childhood, would be followed by the recently added Luminous Mysteries, which detail Jesus' public ministry. Marking his twenty-fourth anniversary as Pope, Supreme Pontiff John Paul II declared the year from October 2002 to October 2003 as "The Year of the Rosary," and requested the addition of five more events in the life of Christ to be added to the centuries old Catholic tradition of the rosary so that the prayer "could broaden to include *the mysteries of Christ's public ministry between his Baptism and his Passion.*" Pope John Paul II asserted that it was important to include these events because, "It is during the years of his public ministry that *the mystery of Christ is most evidently a mystery of light.*" The pope wanted followers to reflect upon "certain particularly significant moments in his public ministry" before reflecting on the suffering of Jesus before His death.[157] The events that Pope John Paul II chose as "particularly significant" demonstrate the divine nature of Jesus *within* His own lifetime, making apparent that Jesus

157 For Pope John Paul II's complete address on the subject of the addition of the Luminous Mysteries, see *Apostolic Letter Rosarium Virginis of the Supreme Pontiff*, 16 October 2002, where Pope John Paul II addresses the Bishops, Clergy and Catholic Congregation on the subject of the Holy Rosary.

was indeed a super-conscious human being *before* His death and Resurrection. An alteration of the prayer which had remained for the most part unchanged since 1569, when Pope Pius V is credited with standardizing the method of praying the rosary, was a surprise to many. The pope's bold move of altering the rosary in 2002 risked agitating the more conventional members of the Catholic Church. BBC reporter Peter Gould explains "that tinkering with the wording would have risked offending traditionalists." Many Catholics were contented with their daily devotion as it was, and were therefore somewhat resistant to the change of a prayer that had remained constant for generations. However, Pope John Paul II was well aware of the disparity this change could have caused, and therefore left the decision of whether or not followers chose to reflect upon the newly added Luminous Mysteries to each congregation.[158] In a time of uncertainty among the congregation, the versatility of the rosary allowed this prayer to continue as a prominent Catholic icon. This new series of meditations added includes reflection on the Baptism of Jesus by His cousin John the Baptist in the River Jordan, which according to Catholic faith, marks the beginning to Jesus' public ministry; the Wedding Feast of Cana, where Jesus is said to have performed His first miracle by turning water into wine; The Proclamation of the Kingdom of God, where Jesus preaches of the Kingdom of Heaven; The

---

158 For the complete article about Pope John Paul II's addition to the rosary, see *BBC News World Edition*, 16 October 2002, "The Pope's Belief in Daily Prayer."

Transfiguration, where Jesus took the apostles to a mountain top to pray and physically transformed in their presence; and the Last Supper, where Jesus gives the Holy Eucharist for the first time.[159] In the next set of reflections, the Sorrowful Mysteries, one would meditate upon Christ's Passion or the proceedings that led up to the Crucifixion of Jesus.

One contemplates the events of the Sorrowful Mysteries on Tuesdays and Fridays. This series of meditations includes the Agony in the Garden, where Jesus prayed in the Garden of Gethsemane shortly before His death and was betrayed by Judas; the Scourging at the Pillar, where Jesus is publicly beaten before a crowd; the Crowning with Thorns, where Jesus is mockingly crowned as the King of the Jews, yet not with a crown of gold and jewels as would be more common for a king at the time, but with a crown of thorns; the Carrying of the Cross, where Jesus carries His cross to the hill of Calvary, sometimes referred to as Golgotha, the hill on which He was crucified; and the Crucifixion, where Jesus is nailed to the cross and dies. The Glorious Mysteries are reflected upon on Wednesdays and Sundays and include, the Resurrection, where Jesus rose three days after He died; the Ascension of Jesus, where Jesus physically rose to heaven, body and soul; the Descent of the Holy Spirit, where the Holy Spirit appears to the Apostles, inspiring them to publicly preach His mission; the Assumption of Mary, where the Virgin Mary is taken body and soul to heaven, an occurrence that according

---

159 Catholics consider the Last Supper to be the First Holy Eucharist.

to Catholic tradition, is unique to any other human being that has ever existed; and the Coronation, where the Virgin Mary is crowned queen of Heaven, again marking the Virgin Mary's special status in the eyes of the Catholic Church.[160] This pattern of mysteries most commonly recited by Catholics is called the Dominican rosary, bearing the name derived from the tradition of the rosary originating with St. Dominic. However, the Cistercians, an order founded by St. Robert, the Abbot of Molesme who felt that other orders of priests were becoming too relaxed with their rituals, reflected on a different series of events when they prayed the rosary. The difference between the Cistercian and the Dominican rosary helps further demonstrate the metamorphosis of the Catholic rosary, including the shift in the overall focus of the prayer from a Marian devotion, a dedication to the Virgin Mary, to a series of reflections centered on the life of Christ.[161]

The Cistercian rosary places a greater emphasis on the life of the Virgin Mary than the Dominican tradition, setting the Annunciation of Mary as the eighth mystery - the first seven being centered on the life of the Virgin Mary as well. In contrast, the Dominican rosary sets the Annunciation of Mary as the first meditation, bypassing the other seven events in the life of the Virgin Mary; thus transforming the rosary

---

160 For a complete description of the mysteries of the rosary, see *The Catholic Encyclopedia*.
161 The *Catholic Encyclopedia* explains that St. Robert began the Cistercian order in order to return priests and religious to the teachings of St. Benedict. St. Benedict, a monk born in the latter fifth century, was renowned for his spiritual wisdom and is credited with being the founder of monasticism in the Catholic Church, a practice that requires one to live under strict religious vows.

from a prayer focused on the Virgin Mary, to a Christ centered reflection. The shift in the overall focus of the rosary could perhaps be explained by the fact that many of the core beliefs of Catholicism revolve around the Eucharist, the belief that the bread and wine offered at the Catholic mass become the actual Body and Blood of Jesus. This belief is reflected in the words said as recipient receives the Eucharist at a Catholic Mass. As a Eucharistic minister[162] gives the Eucharist to the recipient, he or she says to the recipient, "Body of Christ." St. Thomas Aquinas, author of the *Summa Theologica,* one of the most complete works explaining Catholic doctrine, explains, "Since Christ's true body is in this sacrament . ... It must be said then that it begins to be there by conversion of the substance of bread into itself."[163] Yet it should be noted that though the Catholic faith is centered on the Eucharist, that mystery was not added until 2002, the year in which Pope John Paul II added the Luminous Mysteries. In contrast to the Cistercian tradition, the Dominican rosary chronicles the life of Christ from the Virgin Mary's womb to His resurrection.

In the Dominican rosary, any events involving the Virgin Mary relate to her role in the life of Christ, making the meditations more of an "epic story," where each mystery

162 A Eucharistic minister is one who distributes the Eucharist to the congregation after it has been consecrated by the priest.
163 *The Summa Theologica of St. Thomas Aquinas,* trans. Fathers of the English Dominican Province, (Second and Revised Edition: 1920). The *Catholic Encyclopedia* explains that the Eucharist is the actual presence of Jesus in the bread and wine. At a Catholic mass, the priest prays over the bread and wine, and through the process of transubstantiation, the bread and wine become transformed into the Body and Blood of Jesus.

leads into the next.[164] The story-like nature of the Dominican mysteries made the meditations easier to remember that the Cistercian rosary. With the compact, portable nature of the beads and the simplicity of the prayers and meditations, people could recite the rosary anywhere or anytime, whether at home, working in the field, or on a long journey.[165] The ease of remembering the mysteries of the Dominican rosary, coupled with the compactness and portability of the string of beads, would help the rosary become a lasting icon of Catholic identity. But one must ponder why Roche, who lived in the mid fifteenth century, would trace the rosary to St. Dominic and the defeat of the Albigensians three centuries earlier. Aside from trying to foster the Dominican reputation, part of the answer may be found in the events that surrounded the establishment of the Papal Inquisition, an organization established in the mid thirteenth century, which the *Catholic Encyclopedia* defines as, "a special ecclesiastical institutional for combating or suppressing heresy."[166]

The century following St. Dominic's defeat of the Albigensians saw the establishment of the Papal Inquisition. With the threat that the Albigensians posed to the Catholic Church during the twelfth century, along with the spread of Islam, the Catholic Church saw the need for the institution of the Inquisition because in light of the possibility of losing members of the church, "the positive suppression of heresy by

164 Anne Winston, "Tracing the Origins of the Rosary: German Vernacular Texts," 628-632.
165 Ibid.
166 *The Catholic Encyclopedia*, Volume VIII, 1910, s.v. "Inquisition."

ecclesiastical and civil authority in Christian society is as old as the Church."[167] By the time the Inquisition was established, the Crusades, a series of Holy Wars waged by the Catholic Church with the purpose of taking back the Holy Land from the Muslims, had been underway for over 150 years. Though the Crusades were mostly fought against Muslims, at times the Catholic Church would also combat Christians, as in the case of the Albigensians. The establishment of a formal Inquisition justified the "suppression of heresies," either Christian or Muslim, thereby legitimizing the Church's actions.[168] By using St. Dominic and his defeat of the Albigensians as an example, leaders of the Catholic Church such as Roche helped prove that Catholicism was the true faith. By continually propagating the power of the rosary for the next several hundred years, the papacy was able to maintain dedication to the Catholic Church. Pope Leo XIII declared, "The Rosary was instituted chiefly to implore the protection of the Mother of God against the enemies of the Catholic Church, and, as everyone knows, it has often been most effectual in delivering the Church from calamities." Pope Benedict XV, Supreme Pontiff from 1914-1922, avowed, "In the struggle against the Albigensian heretics who uttered horrible blasphemies as they contested all the truths of faith … Dominic defended the sacredness of these dogmas with all his strength, and implored the help of the Virgin Mother by

---

167 Ibid.
168 See *Catholic Encyclopedia*, s.v. 'Inquisition,' for a complete description of the Inquisition and its role in the church.

addressing this invocation to her very frequently: 'Grant that I may praise you, O holy Virgin; give me strength against your enemies'."[169] By using St. Dominic's victory as an example, the Catholic Church made it clear that anyone who spoke against its teachings would be defeated; thus anyone who was in contradiction of Catholic doctrine could indeed suffer the wrath of the Inquisition. As Roche began his proliferation of the rosary in the fifteenth century with the legend of St. Dominic, Father Jacob Sprenger, another member of the Dominican order, would join the fight against heretics as well. In 1475, Father Sprenger founded the Rosary Confraternity, an organization that required one to register his or her name in a book and promise to recite the rosary on a daily basis. These registers were kept yearly, perhaps being used as a means to keep track of the Catholics in the areas which Confraternities were present. Later on, Jacob Sprenger would also write the preface and co-author *The Malleus Malificarum*, which translates to *The Witches Hammer*. For almost 300 years, this book would serve as a manual for the Inquisition on how to locate, torture, and kill a witch, with chapters such as "The Method of Passing Sentence upon one who hath Confessed to Heresy but is not Penitent." At the time of Roche and Sprenger, allegations of witchcraft were rampant throughout parts of Europe, especially among women, with many being tried, convicted and executed as witches, for practices such

169 Pope Paul VI, Pope John XXIII, Pope Leo XIII. *Seventeen Papal Documents on the Rosary,* 103. Rev. Paul Oligny, trans. *Papal Teachings: The Holy Rosary* (Boston: The Daughters of St. Paul, 1980) 164.

as using certain herbs to reduce the pain of childbirth. On 5 December 1484, Pope Innocent VIII issued a special bull, or church document, against witchcraft. This bull coupled with Sprenger's book would lead to mass hysteria throughout Europe, ending with the death of many women, some estimates as high as one million, who were allegedly guilty of witchcraft.[170] Perhaps if a person's name was not on the list of the Confraternity, that individual could be a potential target of the Inquisition and placed under suspicion of heresy or witchcraft, an accusation that seemed to fall upon women more than men as a result of Sprenger's book. The model of St. Dominic and the rosary perhaps gave women a sort of protection from the witchcraft allegations that were so prevalent at the time. By practicing Catholic devotions such as the rosary, women might have been able to avoid being accused of sorcery. The example of the Virgin Mary gave women a model of piety to strive for. Later, in 1892, Pope Gregory XIII would remind followers, "In Mary, we see how a truly good and provident God has established for us a most suitable example for every virtue . ... We strive with greater confidence to imitate her."[171] Propagation of the rosary through the development of confraternities helped attract many followers, since being a member of a Confraternity had many benefits.

---

170 See *Catholic Encyclopedia*. s.v. Pope Innocent VIII.
171 Pope Paul VI, Pope John XXIII, Pope Leo XIII. *Seventeen Papal Documents on the Rosary*. 118-119.

Joining a Confraternity was simple; protocol required one to sign his or her name in a register and pledge to recite the rosary. Being a member of a large praying community appealed to so many because members had the ability to offer prayers on behalf of the deceased as well as themselves.[172] In doing so, one could reduce his or her time or the time of a departed family member in purgatory, which according to Catholic belief, is a condition of temporal punishment; those who have died with unresolved sins remain in an intermediary state for a period of time prior to entering heaven. Prayers on behalf of those in purgatory allegedly reduced one's time there, a practice commonly termed 'an indulgence.'[173] The relief from purgatory in addition to the almost open membership policy of rosary confraternities appealed to many, and within the first seven years of the establishment of Sprenger's Confraternity in Cologne in 1475, membership expanded to over 100,000, and Confraternities began appearing in other parts of Western Europe as well.[174] Such rapid growth can be attributed to the fact that Confraternities had few restrictions; there were no fees to join and everyone was welcome regardless of gender or social class. Other rituals of the Catholic Church, such as the Eucharist, excluded women. Priests were required to perform the blessings over the Bread and Wine of the Eucharist, and since women were forbidden to be priests in the Catholic

---

172 Winston-Allen, *Stories of the Rose.* 111-120.
173 The granting of indulgences continued in the Catholic Church throughout history. In 1832, Pope Gregory XVI issued his Apostolic Letter *Benedicentes*, which discussed how God found the rosary a pleasing dedication, therefore was allowing it as a form of an indulgence.
174 Winston-Allen, Ibid. 4.

Church, as they still are today, perhaps many women found consolation in the fact that they could fully participate in the rosary without the need for a priest acting as a liaison. Little did many women know that the appealing nature that caused them to join the Confraternity to begin with, might have been a way to keep watch over their activity within the Catholic Church. Joining the group that seemed to welcome them with open arms just might have saved their lives. Though both Sprenger and Roche collaborated to keep as many people faithful to the church as possible by advocating devotion to the rosary, they did not agree on using the word *rosary* to represent this dedication, perhaps due to the pagan origin of the symbolism of the rose.

When Alan de la Roche began promoting the rosary in the fifteenth century, he was opposed to using the term *rosary* because he felt the term had "profane associations attached to the rose chaplet" and "worldly connotations." From Ancient Rome to the Middle Ages, a garden cosseted by a rose hedge had been an "ideal" place for romantic encounters. Suitors would give a rose circlet to the girls they wished to flatter, as it was said to denote reverence and respect.[175] Roche's resistance to a term that signified reverence and respect does not seem rational; could there be another reason why Roche was so

---

175 Historians continually debate the dates of the time period referred to as the Middle Ages or Medieval Europe. Some place it as early as 500 AD to as late as 1500 AD. However, the *Catholic Encyclopedia* begins the Middle Ages with the fall of the Roman Empire in 375 AD and ends it with Renaissance Italy in the mid fifteenth century. Bauman, "A Rosary Picture with a View of the Park of the Ducal Palace in Brussels, Possibly by Goswijn van der Weyden," 150.

steadfast about using Dominic of Prussia's term 'Psalter' instead of the term 'rosary'? Perhaps Roche was well aware of the ancient pagan correlation of the rose to Venus, therefore was wary about associating the rose with the Virgin Mary. The ancient Romans celebrated the 'rosalia,' a spring festival that honored the dead. However, even after Constantine the Great Christianized the Roman Empire, a variation of the festival continued into the Middle Ages; the festival was held in May when the roses began to bloom, and roses were worn as headdresses at the celebration. This tradition would further be Christianized as well, giving Catholics the Feast of the May Crowning, a celebration that still continues in the present day. At the May Crowning, the Virgin Mary is honored by Catholic congregations who gather to pray the rosary as a community, and a statue in the likeness of the Virgin Mary is crowned with the circlet of roses. Another example of the translation of pagan to Christian was the Ancient Greek's use of the red rose to symbolize the pierced foot of Aphrodite, the ancient Greek goddess who would later become synonymous with the Ancient Roman goddess Venus. Later, in the book of Wisdom of the Catholic *Bible,* this image would translate to 'The Precious Blood of Our Lord.' The Greek Aphrodite and the Roman Venus, both ancient pagan idols, were revered for their wisdom, as well as their love and beauty. The connection of the rose to both Ancient Greek and Roman paganism may have been unsettling for Roche. At the time of Roche and Sprenger, the Renaissance, or rebirth of ancient learning and understanding, was underway. Some of the ideas experiencing revival did not comply with the teachings of the Catholic Church, as much of the knowledge was derived from Ancient Greek and Roman philosophies and culture, including pagan religious beliefs. Perhaps Roche feared that linking the Virgin Mary's dedication to the rose would too closely equate her with pagan rituals. Opposed to this correlation, Roche favored the use of the term "Psalter," the practice formally propagated by Dominic of Prussia. Sprenger, however, insisted on the term 'rosary' because the rose was a symbol already recognized by many.

Perhaps Sprenger rationalized that it might be easier to present people with an already familiar term and transform it into a Christian icon, as opposed to introducing a completely new term such as 'Psalter,' to which the people could not easily relate.[176] Sprenger's approach indeed triumphed, giving Catholics a 'rosary' instead of a 'Psalter of Mary.' It seems a bit ironic that the same man who wrote the manual to seek out witches was sanctioning the connection of a pagan symbol to the Virgin Mary. It is almost as if Sprenger approved the remainder of a certain level of paganism, or at least enough to conveniently translate to Catholicism. Yet practices such as midwifery and treatment of ailments through the knowledge of herbs, perhaps empowered women too much, therefore such activities would cause many to be accused of being witches. If nothing else, Sprenger was a practical man and his plan achieved its desired goal; many remained faithful

176 Winston-Allen, Ibid. 101-108.

to the Catholic teachings by praying the rosary. It seems fitting that the rose, a symbol formerly linked with Venus and Aphrodite, was chosen in Catholic tradition to represent the Virgin Mary. Venus, Aphrodite and the Virgin Mary have very similar characteristics, as each of the three patronesses was revered for her love, beauty and wisdom.

"The rose was known as the flower of Venus to the ancient Romans. Anything spoken *sub rosa*, or 'under the rose' was sacred and was not to be revealed to the uninitiated."[177] The term "uninitiated" refers to those outside the Cult of Venus, an ancient pagan Roman following that was devoted to Venus and actively sanctioned by the Emperors of Rome beginning in the third century BC. The commissioning of the first temple in honor of Venus was instated on 18 August 293 BC. As Rome became Christianized in the third century AD, cathedrals were built in place of the temples dedicated to Venus, putting Mary in her 'Palaces of the Queen of Heaven,' where she was referred to as the 'Rose, Rose-bush, Rose-garland or Mystic Rose.'[178] The rose remains in the form of the rose window, in gothic cathedrals such as Chartres and Notre Dame, both in France. This process of Christianization would endeavor to erase any residual images of the pagan Venus by replacing them with symbols of the Catholic Virgin Mary; however many similarities between the two patronesses remained.

177 Barbara G. Walker, *The Woman's Encyclopedia of Myths and Secrets.* (San Francisco: Harper Collins, 1983) 866-867.
178 Ibid.

In Catholic tradition, the Assumption of Mary, celebrated on 15 August, is very near to the date of the inauguration of the first Temple of Venus. Catholic doctrine teaches that the Assumption of Mary is the Virgin Mary physically ascending to heaven, body and soul, to exist in the presence of God. This incidence is distinctive to the Virgin Mary, having never been experienced by any other human being.[179] At the time of her Assumption, the Virgin Mary's level of awareness was second only to that of her son, and her graces gave her vision superior to that of any other "blessed" person. "She surpassed the other blessed in her knowledge of creatures, particularly in her knowledge of her fellow man."[180] The Assumption of Mary, the fourth Glorious Mystery of the rosary, bears a striking resemblance to the astuteness for which the pagan Venus is revered. Both Venus and the Virgin Mary were venerated for their wisdom; it was a wisdom possessed by few, a wisdom not bestowed upon the "uninitiated." One might argue that the term rosary never truly adopted a new meaning when it became associated with the Virgin Mary as it maintained its original symbolism of wisdom in several respects. The word 'rosary' began to appear in German and Latin vernacular texts in about the thirteenth century. During this time, the German *rosenkranz* and the Latin *rosarium*, often referred to an anthology of texts, and were therefore appropriate terms to use to describe the Virgin Mary and

---

179 Richard P. McBrien, *Catholicism: New Study Edition*, revised and updated. (San Francisco: Harper Collins, 1994), 1101-1102.
180 *New Catholic Encyclopedia* 2nd Ed. (Washington DC: Catholic University of America, Thomson and Gale, 2003).

her wisdom.[181] The wisdom of both the Virgin Mary and Venus could be seen as a kind of metaphorical anthology. So although Sprenger advocated the term rosary based on its familiarity to the public, it can also be argued that the term maintained its original meaning as well. As Venus was admired for her wisdom, so was the Virgin Mary at the time she was called to be with God. The similarities between the two women allowed for a suitable translation of the term rosary into the Catholic belief system, helping Christianize former pagan beliefs.

> Crimson rose of heavenly fragrance ... Beautiful Mary, you raised women to a new dignity. Adam's Fall did not stain thee. Thy body did not know corruption. Peaceful sleep fell upon thee. Thy feet on the wings of angels.

Thomas Aquinas, *Sunday Sermon*

As mentioned earlier, the connection between the rose and the Virgin Mary began with early prominent figures in the Catholic Church. St. Thomas Aquinas, born in 1225, wrote the *Summa Theologica*, a complete work on Catholicism that would be used both to instruct beginners as well as teach the proficient in the Catholic faith. This work would

---

181 Josef Klapper, "Mizellen: Central German texts from Breslauer Handwriting," *Magazine for German Philology*, 47 (1918), 83-87.

eventually earn him the title 'Doctor of the Church' and he often praised the Virgin Mary in sermons, at one time using the phrase, *ave rosa spina careens*, 'a rose without thorns.' Aquinas' reference to the Virgin Mary being 'without thorns' symbolizes the Catholic belief that the Virgin Mary was born without original sin, the sin that all are other born with as a result of Adam and Eve eating the fruit from Forbidden Tree in the Garden of Eden. This seemingly miraculous state at the moment of the Virgin Mary's conception is celebrated yearly in the Catholic faith on 8 December, with the Feast of the Immaculate Conception. Once again the Virgin Mary is distinguished from other human beings, even from the moment she was conceived. Later on Peter Martyr, a member of the Dominican order in the thirteenth century who would eventually become the Inquisitor General of Italy, called the Virgin Mary *ave java mundi rosa*, or 'radiance of the world.' Similarly, St. Catherine of Siena, a Dominican nun who dedicated her life to helping the poor and the sick, praised the Virgin Mary as 'the rose of heaven.'[182] Roche and Sprenger publicized these already formed connections between the Virgin Mary and the rosary at the same time alleged witches threatened the power of the Catholic Church. At this time, the rosary was also a way to satisfy the needs of the people during the continued devastation of the Black Death, a plague transmitted by fleas and carried on the backs of rats

---

182 References to St. Peter Martyr, St. Dominic and St. Catherine of Siena were found at NewAdvent.org, an online compilation of over 11,000 articles from *The Catholic Encyclopedia, Summa Theologica, Church Fathers, Bible*, and *How to Recite the Holy Rosary.*

that eliminated one third of Europe's population during the Middle Ages.

Though the Black Death made its first appearance in Europe in the fourteenth century, the plague periodically swept through the continent in waves, reappearing as late as the seventeenth century. At times, the plague was wiping out as much as one third of entire village populations.[183] Priests were able to offer little relief to the masses for several reasons; the first was that the Black Death knew no profession, many priests also fell victim to the disease therefore leaving some towns without a parish priest. Another reason that priests could not help was because even if they were brave enough to face the sick, knowing the possibility that they could fall ill themselves, there was little assistance they could give without the aid of modern medicine. As a result, people began losing faith in the Catholic Church. The devastation of the plague caused people to desire a way to show their dedication to God without having to go to a priest, who served as an intermediary. At this point in European history, significant portions of society, such as the poor and the peasants, who comprised the majority of the population, were either illiterate or not versed in Latin, the language of the Catholic Church at the time, and could not decipher written prayer books. The rosary provided a solution to all these issues while keeping congregations committed to the Catholic Church; it provided people with a more direct involvement in devotion to God

---

183   Black Death, *The Concise Oxford Dictionary of Archaeology*. (NY: Oxford UP, 2003).

with something they could easily remember and carry with them at all times, and also helped restore their faith in the Catholic Church.[184] The practice of the rosary also helped keep followers faithful to Catholicism during the Protestant Reformation, a movement that began in 1517.

Prior to the Protestant Reformation, some corrupt leaders of the Catholic Church were selling indulgences for money, allowing followers to believe they could reduce their time, or a deceased family member's time in purgatory for a sum of money. Martin Luther, a Catholic monk, publicly opposed the Catholic Church's retailing of indulgences. As a result of Luther's protest, some Catholics completely broke away from Catholic teachings to follow the teachings of Martin Luther, and later became known as Lutherans. In response to the significant loss of its members, the Catholic Church promoted the rosary as an indulgence that would perhaps prevent followers from straying to the newly formed Lutheran sect of Christianity. The Catholic Church was faced with a bit of a dilemma when dealing with the issue of purgatory and indulgences during the Protestant Reformation. The doctrine of purgatory was already cemented in the minds of the people, therefore it would not have been feasible for the Church to renege on this teaching, but as a result of Luther's protest, followers were no longer left with a means to reduce their time there. Since the sale of indulgences was causing the church to lose members, the

---

184 Anne Winston, "Tracing the Origins of the Rosary: German Vernacular Texts,"
    635.

Catholic Church needed to give followers an alternative means to help reduce time in purgatory while preventing the further loss of members. The rosary, already a familiar practice to Catholics, could conveniently fill this need. As late as 1832, popes continued the proliferation of the rosary as an indulgence, including a statement by Pope Gregory XVI in his Apostolic Letter *Benedicentes*, which stated, "We have, therefore, decided to grant papal approval to this beneficial institution and to enrich it with indulgences." Throughout many parts of Europe, the rosary served as a way to keep followers faithful to Catholicism, for now Catholics had a way to reduce their time in purgatory without having to pay corrupt church officials. However, in England the rosary would serve a one of the last remaining links to Catholicism for English Catholics.

In 1559, Queen Elizabeth I of England issued the Second Act of Uniformity, making Protestantism the solitary lawful religion in England.[185] By the 1570s, few Catholic priests remained in England and most Catholic images and vestiges in the country, including churches, were destroyed as a result of Anglicans, the followers of the Church of England, opposing what they considered the ritualistic nature of the Catholic mass. In spite of this hostility, devotion to the

---

185 The first Act of Uniformity was issued by Henry VIII, the founder of the Church of England. After his death, there was a continual power struggle between Catholic and Protestant rulers, such as that of "Bloody Mary" the queen who wanted to take revenge on the Protestants, and therefore sanctioned their slaughter. As a result, devotion to Protestantism grew, defeating Mary's purpose of reinstating Catholicism as the faith of England. When Elizabeth came to the throne, religious turmoil was tearing her country apart, therefore decided to unify England under one faith.

rosary actually increased. Because of its size and portability, the string of beads could be masked effortlessly in a man's pocket or sewn inside a woman's dress, therefore allowing Catholics to continue with devotion to their beliefs. the Act of Uniformity, participation in the Eucharist was impossible in England because of the illegality of Catholic ceremonies. The rosary was one of the only ways for English Catholics to express their devotion to the faith.[186] According to Catholic tradition, lack of participation in the Eucharist would increase one's time in purgatory. As mentioned earlier, praying the rosary would help reduce a person's time there; therefore the prayer of the rosary would help combat the trouble inflicted by the Protestants. A Catholic's time in purgatory was increased because of lack of participation in the Eucharist, but was reduced by the praying the rosary. Its popularity would rise as a means of contesting Protestant actions against the Catholics of England. While the rosary is a devotion to the Virgin Mary, one accesses Christ through her, "the source of all saving grace."[187] On 1 September 1883, Pope Leo XIII explained in his *Supremi Apostolatus,* the connection of the Virgin Mary to her son Jesus when she is prayed to for guidance. "She, who is associated with Him in the work of man's salvation, has favor and power with her Son greater than any other human or angelic creature has ever obtained or ever can obtain."[188]

---

186 Lisa McCain, "Using What is at Hand: English Catholic Reinterpretations of the Rosary, 1559-1642," *The Journal of Religious History* 27 no. 2, (2003): 161-176.
187 Ibid. 162-165.
188 Pope Paul VI, Pope John XXIII, Pope Leo XIII. *Seventeen Papal Documents on the Rosary.* 99.

The rosary allowed devotion to Catholicism to remain strong for Catholics in Elizabethan England. William Eric Brown, author of *John Ogilvie: An Account of his Life and Death with a Translation of Documents Relating thereunto,* explains the fate of John Ogilvie, an English Catholic who was executed for treason. Before his death, Ogilvie bid farewell to his fellow Catholics, taking his rosary with him to the gallows. As a final act, he threw it into the audience where it landed on an observer.[189] Though practicing Catholicism posed its risks, Ogilvie's bold action demonstrated the deep devotion that still existed for some remaining Catholics in Protestant England. Perhaps Ogilvie's last gesture was one of defiance. Since he was already destined for the gallows, possibly he took the risk knowing he had nothing to lose. Another explanation could be that he was, in fact, showing one final act of defiance by stating he was indeed still a faithful Catholic in a country that legally no longer allowed him to be. In England, Catholics attempted to cling to the rosary, as it was one of the last remnants of Catholicism in their Protestant country. However, in other areas of Europe, such as France and Spain where the Catholic Church still had the majority of followers, the rosary was actively endorsed.

As the Protestant Reformation continued to take members of the Catholic Church throughout parts of Europe, Pope Pius V and Pope Gregory XIII sanctioned the Feast of the Holy

---

189 William Eric Brown, *John Ogilvie: An Account of his Life and Death with a Translation of Documents Relating thereunto* (London: Burns, Oates, and Washburn, Ltd. 1925) 144.

Rosary to be celebrated on 7 October 1571. On this date, the Muslims of the Ottoman Turkish Empire made one last attempt to expand their territory as far as Spain, but were defeated in the Battle of Lepanto. This victory over the Turks was credited solely to the Virgin Mary. It was said that as the battle was being fought, the Confraternity of the Rosary was meeting at the Dominican headquarters in Rome, where they recited the rosary with the belief that doing so would lead to the defeat of Muslim enemies.[190] In the midst of the Protestant Reformation, the declaration of the Feast of the Holy Rosary reminded Catholics of their true faith. Perhaps the papacy used the defeat of the Turks as an example, helping show the importance of remaining true to Catholicism in the midst of others straying from traditional Catholic beliefs. With parts of Europe experiencing religious turmoil, both Catholics and Protestants were vying to be the dominant religion. The Feast of the Holy Rosary could have conceivably served as a reminder of the possible fate of non-Catholic believers, that they too would be defeated just as the Turks were, and not just by an army, such as that of Spain which defeated the Ottoman Turks, but by Divine Intervention, much how the Albigensians were allegedly defeated a few centuries earlier when being slaughtered during a Crusade. It is interesting to note that the defeat of the Ottoman Turks a result much like the defeat of the Albigensians of southern France. However, propagation from popes and other church leaders was not

---

190 Joseph Therese Agbasiere, "The Rosary: its history and relevance." *African Ecclesial Review* 30 no. 4 (2006) 250.

the only method used to foster devotion to the Virgin Mary's dedication.   Artists, especially during the Medieval and Renaissance periods played a significant role in the process as well.

Medieval and Renaissance art helped define a universal format for the rosary, allowing it to gain a foothold as well as helping sustain its general recognition. The invention of the printing press in 1452 made vernacular prayer books, books that were printed in the language commonly spoken in a particular area, more readily available. However, other alternatives were needed for the illiterate populace. To help the rosary gain popularity, artists began using it as a focal point for their work. Purgatory was a common theme; those unable to read popes' decrees could see the saving power of the prayers of the rosary beautifully depicted before them. Regardless of what language one spoke, anyone could appreciate their fate portrayed by the vivid images created by artists.

In 1483, one of the most influential depictions of the mysteries of the rosary was created in woodcut for the brotherhood of Ulm by an artist named Dinkmut. No writing accompanied the work; therefore the message could reach all populations, including the poor, illiterate and those of various linguistic backgrounds. As late as the sixteenth century, this visual adaptation of the rosary was more popular than any written version. Pictorial versions of the rosary varied little from region to region, all portraying the mysteries of the Dominican rosary in some way. Each set of mysteries was often represented in a different color. The Joyful Mysteries were represented as a white rosary, detailing the birth and childhood of Jesus. The Sorrowful Mysteries were depicted in red, showing the passion of Christ. The golden rosary represented the Glorious Mysteries, signifying resurrection of Christ and the Assumption of Mary.[191]

In 1488, the Spanish artist Francisco Domenech created a woodcut called *The Fifteen Mysteries and the Virgin of the Rosary*, which now resides in the Metropolitan Museum of Art in New York City. The painting consists of sixteen panels which portray the Joyful, Sorrowful and Glorious Mysteries of the rosary, as well as the Virgin Mary. To the right of the Virgin Mary is a kneeling nobleman who holds a string of rosary beads and three roses. This entire panel is surrounded by a string of rosary beads made from actual roses. The Virgin Mary is holding the Christ child, who is holding one of the flowers that makes up the perimeter of the panel. The nobleman has three assaulters in the wake of him, to which he seems oblivious. Out of his mouth comes a stem of three roses and on his head rests a garland of roses.[192] This panel of the Virgin Mary and the knight gives some insight to the early legends surrounding the rosary. In 1408, Vincent Ferrer, a Dominican preacher, attempted to unite the church leaders of Roman Catholicism and Eastern Orthodoxy, a sect of Christianity that observes the Byzantine Rite and the

---

191 Winston-Allen, *Stories of the Rose.* 8-32.
192 Francesco Domenech, *The Fifteen Mysteries and the Virgin of the Rosary.* Engraving, 1488, The Metropolitan Museum of Art, New York.

Patriarch of Constantinople, as opposed to the papal rule in Rome. Ferrer told of a knight who began to recite the rosary after being captured. When his captors returned, they saw the Virgin Mary along with St. Catherine of Siena, a fourteenth century Dominican nun who had visions at an early age, and Agnes of Eulalia, a fourth century martyr who was murdered under the Roman Emperor Diocletian. Catherine held a plate of roses and Agnes held a needle and thread. As the knight prayed, each 'Hail Mary' produced a rose which the Virgin Mary strung into a crown and placed on the knight's head. The assaulters were astounded and inquired who the lady was; the knight was unaware that anyone else was present. The captors released the knight and decided to convert to Christianity.[193] For the illiterate population, a work that depicted imagery as powerful as this would give a prayer reference and a means of inspiration. The British Museum houses a work very similar to Domenech's which illustrates the mysteries as well.

The accompanying story to this version tells of a chap who would make a circlet of roses each day as a means of expressing his commitment to the Virgin Mary. Later he entered a Catholic monastery where he was unable to execute this daily custom. The leader of his order instructed him to say fifty 'Hail Marys' in place of the physical wreath he was accustomed to making. While traveling in the forest

---

193 Valerio Serra and Boldu, *Libre d'or del rosary a Catalunya* (Barcelona, 1925) 22. See *Catholic Encyclopedia* for detailed explanation of St. Vincent Ferrer, St. Catherine of Siena and St. Agnes.

one day, he stopped to say his prayers and was interrupted when an assemblage of bandits that attempted to kill him. As they approached him, they saw a remarkable lady holding a chaplet; she was taking roses from the mouth of the monk and adding them to her wreath. She completed the wreath, placed it on her head and vanished. The captors had been watching close by and questioned the monk, who was unaware of the presence of the lady. They then set him free and converted, realizing the miracle they had witnessed.[194] Both versions of the legend are strikingly similar, though the paintings describing the stories are housed on different continents. Another illustration of the rosary, credited to Flemish artist Goswijn van der Weyden, also resides in the Metropolitan Museum of Art. This version is comparable to the already mentioned adaptations displayed at the Metropolitan Museum of Art and the British Museum.

In 1984, the Metropolitan Museum of Art received a series of sixteen paintings representing the fifteen mysteries of the rosary. The final panel shows the Virgin Mary in the center with St. Dominic on the left and a pope, an emperor and a king on the right, all facing a kneeling knight with three assailants behind him. The Virgin Mary is carrying the Christ child, who holds a rosary made of physical roses, fifty white roses representing the 'Hail Marys' and five red roses representing the 'Glory Bes' and 'Our Fathers.' This painting is thought to have been commissioned by a member of the Hapsburg

---

194 Bauman, "A Rosary Picture with a View of the Park of the Ducal Palace in Brussels, Possibly by Goswijn van der Weyden," 24:141.

kin, a royal German family whose members ruled throughout Europe from the late Middle Ages until the twentieth century.[195] The similarities between the works of Domenech, van der Weyden and the work in The British Museum give modern day devotees a glimpse of the fundamental aspects of the rosary. The fifteen mysteries are integral to the devotion, and therefore appear in all three versions. The Virgin Mary is portrayed in each of them, which reinforces her significance to the Catholic faith. The fact that the rosaries are made of actual roses links the Virgin Mary to the wisdom for which she is revered. Although the above mentioned works of art show the Catholic link of the Virgin Mary to the rosary, a print, dated 1581, is housed in the New York Public Library and presents the Virgin Mary much in the same way Boticelli presents Venus in the fifteenth century work, *The Birth of Venus.*

Roman mythology tells of the goddess Venus being born from the sea as a result of her angry mother Gaia, the goddess of Mother Earth, castrating her father Uranus, the god of the Sky and Heavens, and throwing his genitalia in the sea. As a result, her father's semen mixed with the foam of the sea and Venus was born as an adult emerging from a scallop shell and arrived on the shores of Cyprus. One should note that the Virgin Mary, as mentioned earlier, also has a special circumstance surrounding her birth. Though the Virgin Mary did not emerge from the sea, her state of

195 Ibid. 135.

freedom from original sin at the time of her birth sets her apart from any other human being. Venus was unique at the time of her birth because of her being born as an adult from the sea. In Botticelli's painting, *The Birth of Venus*, the red-haired goddess appears to emerge from a giant shell and is surrounded by angel-like figures. Pink flowers, which appear to be roses, blow in the wind around her. An unaccredited print in the Spencer Collection of the New York Public Library titled *Prayers and Feast Days*, demonstrates the Virgin Mary emerging from nature in a similar manner. A red-headed Virgin Mary holds her son in her left hand, but appears to have no lower extremities; instead she emerges from a giant rose. In her right hand, she holds a small rose toward which her naked infant reaches. A subtle glow illuminates her head. This sixteenth century work physically connects the Virgin Mary to the ancient symbol of wisdom, the rose. The *Catholic Encyclopedia* published in 1911 mentions that "Venus too often masquerades as the Madonna."[196] Madonna is another term often used to describe the Virgin Mary in the Catholic tradition. The second half of *Prayers and Feast Days* consists of a woman with seven children, all joining hands to form a circle around her. All faces are looking at the audience and in the background, a brick wall seems to enclose the group. The woman is flanked on both sides with roses.[197] Renowned for her knowledge, the Virgin Mary is surrounded by roses, the

196 *The Catholic Encyclopedia*, Volume XII, 1911, s.v. "Venus."
197 *Prayers and Feast days*, 1581, NY Public Library, The Spencer Collection, New York.

flowers synonymous with the wisdom of Venus. Revering the
Virgin Mary's knowledge and its importance to the Catholic
tradition, followers flocked to the Virgin Mary as children
would to a teacher or a mother, as seen in *Prayers and Feast
Days*.

Another unaccredited work at the New York Public
Library, titled *Noble Milanaise*, dates from 1860-1861 and
depicts a woman from the 1300s. While nobles in Medieval
Europe were traditionally adorned in plush fabrics to
display their affluence, this woman is veiled and layered in
very simple fabrics. She wears a basic red tunic tied under
her bosom; her light blue shawl, draped around her, reaches
the floor. She has a white veil, which follows her hair line,
keeping her tresses covered, but allowing her face to show.
Her eyes are cast to the ground and gaze to the right,
although it is not clear what she is looking at, as she is the
only element present in this work. Her arms are folded and
she is holding a rosary in her left hand. Her appearance is
not one of wealth or distinguished bloodline, yet she is given
the title "noble" by the unknown artist.[198] She is a rendition
similar to many depictions of The Virgin Mary in Medieval
and Renaissance art, such as Jean Bourdichon's *Miniature of
Annunciation*, another print housed at the New York Public
Library.[199] The Catholic Church often preached that there
was no better model of humanity than the Virgin Mary, the

198　*Noble Milanese*, 1860-1861, NY Public Library, Mid-Manhattan Picture Col-
　　lection, New York.
199　Jean Bourdichon, *Miniature of Annunciation*, 1457?-1521?, NY Public Library, Medi-
　　eval and Renaissance Illuminated Manuscripts from Western Europe, New York.

essence of *spiritual* nobility. By labeling someone dressed in the described manner as a noble, artists suggested to women to imitate this humble image. However, while visual art would serve an illiterate population in the Middle Ages and Renaissance Europe, more recent decrees from popes would help "foster" devotion among a more literate population.[200] Reverend Paul J. Oligny translated a letter written by Pope Gregory XVI in 1832 that avowed, "The rosary is not only right because of the words, but also because it brings people closer together as they pray it, therefore God finds it favorable."[201] Pope Leo XIII, who would later become known as the 'Pope of the Rosary,' went on to issue nine encyclicals or official letters, specifically referring to the rosary. On 1 September 1883, he issued his Encyclical *Supremi Apostolatus* which declared:

> We consider that there can be no surer and more efficacious means to this end than by obtaining through devotion and piety in favor of the Virgin Mary, the Mother of God, the guardian of our peace and the minister to us if heavenly grace, who is placed on the highest summit of power and glory in heaven in order that she may bestow the help of her patronage on men who through so many labors and dangers are striving to reach the eternal city. (Pope Leo XIII, 1883)

---

200 *The Holy Rosary: Selected and Arranged by the Benedictine Monks of Solemes*, trans. Rev. Paul J. Oligny, OFM, 31.

201 Ibid.

Leo XIII goes on to remind his readers of the long-standing tradition of the rosary in the Catholic faith, labeling the heretics of both the Albigensian Crusade and Battle of Lepanto as the "dangers." As recently as 1955, Pope Pius XII addressed an international group of rose growers in Italy, commenting on the importance of the symbolism of the rose in Catholicism. He explained that it was the first Christians, not Roche alone, who rejected the rose because it was representative of "a life which they abhorred."[202] The rose, once allegedly rejected by Christians for being too pagan, would later adorn some of the most famous Medieval Cathedrals, such as Notre Dame, in the form of the stained glass rose window. Pius XII continues to explain that the rose's connection to the Virgin Mary stems from St. Bernadette's apparitions. St. Bernadette of France was said to have a total of eighteen visions of the Virgin Mary at Lourdes, France. Many times, there were others around but Bernadette was the only one who could see the visions.[203] According to Pius XII, "It is fitting that the most beautiful of flowers is an offering to the most beautiful of creature."[204] While Pius XII would actively remind the Catholic congregation of significance of the rose throughout the history of their faith, Pius XII's successor, Pope John XXIII, would call an entire council that would alter the status of the rosary.

---

202 Pope Pius XII, "To an International Group of Rose Growers in Rome," *The Pope Speaks* 2 (Summer 1955): 133-135.

203 Today a Cathedral stands, said to be instructed by the Virgin Mary, and Lourdes, France remains a popular site for Catholic pilgrims.

204 Pope Pius XII, "To an International Group of Rose Growers in Rome," 134.

Earlier history shows the rosary being used at times to fulfil the needs of the illiterate population as well as to maintain the faith of Catholics when the church was in danger of losing followers. But during the reign of Pope John XXIII from 1958 to 1963, illiteracy was no longer a problem facing the Catholic Church; followers were, however, looking for a more active role in church rituals. The church officials of John XXIII's time sought to embrace the problems of the contemporary world. John XXIII called the Second Vatican Council in 1962, which helped church leaders address the spiritual needs of twentieth century Catholics.[205] Catholic doctrine was reexamined in order to better suit twentieth century congregations. Prior to this Council, Catholics were not actively encouraged to read the *Bible,* as it was believed that lay people could not fully understand the meaning of Scripture. The term 'lay people', or 'laity,' refers to those who are not of a particular religious order the way priests and nuns are, but still identify themselves as adhering to the teachings of a particular religion. Prior to the Council, it was believed that the laity needed a priest to interpret the Gospels in a Homily, a sermon given by the priest at a Catholic Mass in which the priest relates the Scripture reading to everyday life. However, the Second Vatican Council realized it was beneficial to actively encourage people to read Scripture outside of the mass as well, as the general congregation was literate and much more educated then the medieval followers who would have been unable to read the Scriptures. Though

---

205 See *The Catholic Encyclopedia* for a complete history and description of the Second Vatican Council.

the Homily still remains a part of the Catholic Mass today, the need for the rosary as an alternate prayer form was no longer as urgent after the Second Vatican Council as it was during the time of the Black Plague and the Protestant Reformation. People now began reading Scripture as a way to show their devotion to God as well. That is not to say that the Second Vatican Council *discouraged* dedication to the rosary, far from it. The Council simply offered the general congregation another alternative of devotion outside the formal Catholic Mass. With one billion Catholics in the world today, devotion to the rosary remains, but the desire for an alternate prayer form is not as vital as it was during the Middle Ages.[206] Many remaining devotees were raised before the assembly of the Second Vatican Council, with the rosary being a vital component of their daily lives. For this generation of followers, the rosary was already an integral part of their daily ritual; therefore they were comfortable with the practice and decided to continue with their devotion to the rosary. Catholics worldwide use the same strings of beads and reflect on the same mysteries when they are praying the rosary. However, there are some slight variations to the overall ritual. When the mysteries are read as well as which prayer ends the devotion depends the country in which one prays. As in the time of Roche and Sprenger, the flexibility of the devotion to the rosary continues to appeal to many, helping it remain a lasting Catholic icon.

206 McBrien, *Catholicism.* 1067-1071.

In the United States, the rosary ends with the 'Hail Holy Queen.' In Ireland, the rosary ends with the 'Litany of the Blessed Virgin.' In Uganda it ends with the 'Memorare.'[207] Though the ending of the devotion varies from one country to another, the central prayers of the rosary and content of the mysteries remain the same, as they are the core of the rosary dedication. Though there are some minor differences, the content of the mysteries remains unchanged. The mysteries, taken straight from the events of the <u>Bible</u>, are the crux of both the rosary and the beliefs of Catholicism, unwavering throughout centuries of the Catholic tradition.[208] When read in the United States, the leader of the prayer announces the mystery before each decade of 'Hail Marys' begins. In German-speaking countries, however, the mystery is said after the line in the 'Hail Mary,' "thy womb Jesus." For example, the fifth Sorrowful Mystery would read, "thy womb Jesus, who was crucified for us."[209] Just as the paintings created by different artists and housed in various museums maintained the central elements of the Catholic rosary in a consistent format, so do devotees around the world, who continue to recite the rosary, preserving it as lasting symbol of Catholic faith.

207 See *The Catholic Encyclopedia*. The 'Litany of the Blessed Virgin' is one of the many prayers in honor of Mary. The 'Memorare' is often attributed to the French priest Claude Bernard but dates before his time. All three prayers are prayers said to honor the Blessed Virgin.
208 Agbasiere, "The Rosary: Its History and Relevance." 30:250.
209 *New Catholic Encyclopedia* Volume XII, (NY: McGraw Hill, 1967) 667.

For Catholics, the 1682 year old tradition of the rosary, which began with Constantine the Great's Christianization of Rome, has incessantly served as a devotion to and a model of the piety of the Virgin Mary. Though the *official* origin of the rosary remains a topic of a discussion, the significance of the rosary's connection to the rose continues to linger. While Alan de la Roche attributed the entire tradition of the rosary to St. Dominic, the significance of the rose actually pre-dates Christianity, as seen through this essay, with its roots lying in the earlier pagan Roman goddess Venus. However, the formal recognition and spread of the rosary can be credited to the Dominican order, where members of the order such as St. Catherine of Siena, St. Peter Martyr and Alan de la Roche would actively preach the importance of the dedication to the rosary. The most active endorsement occurred during times of both political and religious turmoil, such as the Inquisition, the Black Plague, and the Protestant Reformation, all periods in which the Catholic Church was in danger of losing members of its congregation. Thus the rosary has been used in the Catholic tradition in more ways than one, helping demonstrate the endless flexibility of the rose as a timeless symbol. Aiding in this diffusion of the rosary was Medieval and Renaissance art, which helped establish a universal format for the rosary devotion. Most recently five more meditations were added to this centuries old tradition, with the addition of the late Pope John Paul II's Luminous Mysteries in 2002, showing the continual flexibility of the rosary right up to the present day. And yet, in all endorsements of this popular

Catholic prayer, the uniqueness of the Virgin Mary is always remembered. Entire devotions to the Virgin Mary such as the rosary, prayers like 'Our Lady of Guadalupe' that end with the line, "Mystical Rose, pray for us," and works of art such as Domenech's *Fifteen Mysteries and the Virgin of the Rosary*, continually embrace the symbolism of the rose and the ancient wisdom possessed by the Virgin Mary.

6

The Rose and Astrology

By Montgomery Taylor

Astrology is, first of all, a language of symbols. Although vulnerable to misinterpretation, those symbols communicate truth beyond the capability of words and when carefully reflected upon can help us understand who we truly are. The following essay is intended to explain, in simple terms, how the field of astrology or the study of the influence of heavenly bodies on human affairs is experiencing a renaissance and rebirth as a tool for psychological insight and for the forecasting of events and conditions. After suffering the condemnation of man's primitive understandings of the material universe, the great Swiss psychologist Carl Gustav Jung, 1875 – 1961, did much to restore the value of astrology as a means of more clearly delving into the individual and collective subconscious of man. Many of the world's most esteemed astrologers cite the work of Carl Jung, pointing out that symbolic realities are just as important to human nature as are physical realities, maybe more important.[210] Ironically, Jung, as a master astrologer and often did astrological charts for his most problematic patients with great results. Astrology, according to Jung, was actually the oldest form of psychology because through astrology linear mental thinking can be aligned with intuition and emotional patterns.

An effective analogy that can help in the understanding of astrology is to see planets in the same manner as we view satellites. Planets are indeed satellites that orbit the sun along with the earth. When we view planets from the earth's perspective, we are actualizing what is called geocentric astrology, that is, planetary patterns in the heavens as viewed from earth. Just as our man-made satellites have specialties such as weather, communications, military, etc. so do planets have specialties in terms of abstract cosmic energy that affect earthly events, including the human psyche. These energies are best represented in terms of human understanding as archetypes.

Jung put forth the concepts that the gods and goddesses of ancient mythology throughout the world were symbolic in their nature and then shared by all humanity. Jung likewise postulated the existence of a level of consciousness shared by all humanity, which he called the "collective unconscious".[211] Myths from every culture deal with the same subjects of mortality and immortality. The stories vary, but the messages

210 Ariel Guttman and Kenneth Johnson, *Mythic Astrology* (St. Paul, Minnesota: Llewellyn Publications, 1993) 2.
211 Ibid. 3.

are consistent giving human beings a perspective of their relationship to universal consciousness. The whole point of astrology is to bring the unconscious information to conscious awareness.

Finally, true astrology, as opposed to the entertainment found in popular newspapers, is being given a place of respect in universities and schools of higher learning. The world's leading astrologers have founded professional research organizations such as the National Council of Geocosmic Research and others to assure the highest integrity and training of its practitioner members. Many consider astrology to be man's earliest attempt to explore psychological and spiritual matters. After all, it was three astrologers that predicted the birth of Jesus! And to this day, the celebration of Easter is calculated as "the first Sunday after the first full moon, after the sun has entered the constellation of Aries." This explains why Easter is celebrated on a different date every year, but the other holidays remain the same.

Having a horoscope or natal chart calculated for the average person is a very recent development of the last century. Throughout history to the present day, royalty and world leaders from Queen Elizabeth I of England to presidents of the United States have taken astrological advice into consideration. In fact, since most of the lower classes had no idea of the exact time and date of their birth, horoscopes were usually cast only for royalty and heads of state or members of the aristocracy. In the secret writings of Freemasonry, it is a well-known fact that Benjamin Franklin was a master astrologer and actually used the science of astrology in choosing the date for the signing of the Declaration of Independence.[212]

In the mid-20th century, a team of expert French researchers in statistics, Michel and Francoise Gauquelin, spent years compiling data with a view of disproving astrology. However, to their great surprise, their research resulted in an unquestionable validation of this ancient science. Thousands of birth dates and related personal information fed into their computer convinced the Gauquelins that an uncanny consistency existed, that the movements of planetary bodies played a major role in identifying character traits, that those movements would lead a person to have distinct identifying traits and correlate eventually to certain career paths. As a result of their research, British astronomer Percy Seymour developed the theory of celestial magnetism to account for the influence of planets on human nature.[213]

Metaphysics and Quantum Physics

We may look at metaphysics as a poetic expression of quantum physics, where spirit and science validate each other in different languages. During these times of expanding awareness and curiosity, many polls reveal a growing

---

212 Robert Hieronimus, Ph.D. *America's Secret Destiny: Spiritual Vision and the Founding of a Nation*. (Rochester, VT: Destiny Books, 1998) 37, 44.
213 Ariel Guttman and Kenneth Johnson, *Mythic Astrology*. 1.

interest in the mystical and unexplained.[214] If we approach spiritual experience with a perspective that is solely based on skepticism and doubt, nothing can be learned, and we doom ourselves to look at higher Universal awareness as nothing but a myth. This attitude became prevalent and imbedded during the seventeenth century, a period historically referred to as the "Age of Reason"; humankind wanted to see the world through the eyes of pure mental logic and linear reasoning. We quite understandably began the modern age with an overly simplistic and materialistic view of our universe.[215] Granted, much of the pure wisdom and search for an understanding of humanity's relationship to the universe as found in the ancient mystery schools left a legacy that was distorted and devolved during the medieval period. It became corrupted, misinterpreted, wrongly translated and was a victim of charlatans, religious zealots and condemnation by totalitarian religions masking as spirituality.[216]

Consider a rather simplified but effective explanation of how our rich and varied heritage of myth can be compared to a hypothetical but not far-fetched example of a modern-day airplane pilot encountering a Stone Age culture. Imagine your helicopter or small engine plane making an emergency landing in the darkest depths of the Amazon jungle. The local natives look at you with fear and wonder. Your appearance is different from theirs and you landed in their midst in a

---

214 James Redfield, *The Celestine Vision* (NY: Warner Books, 1997) xix.
215 Ibid. xxiii.
216 Ibid.

great bird or flying dragon from the heavens. They have never seen anything like you before. How will you communicate with them? How will you explain your technology? Then your cell phone rings, complete with talking pictures. You are telling the caller that you have landed safely, and you give your location so that you can be rescued. In the meantime, you try to explain your presence and your technology. But how can you do that? You have to invent a story that explains aerodynamics and reassures the natives that you do not have a god or a disincarnate slave trapped in your cell phone. They marvel at their new understanding that you give them in the form of your own allegorical myth, which tries valiantly to represent the facts that are natural and normal to you but are otherworldly to them. So it is with our astrological myths. The ancients of every culture could not help but notice that certain circumstances and events occurred with regularity whenever the sun, moon, or certain planets were in places that formed specific relationships to each other when observed from the earth.

And thus were born the "archetypes" – the characters that are defined in the dictionary as "a first form or model; the original pattern or model after which a thing is made."[217] The leading characters, or archetypes, of our astrological myths were merely man's attempt to put an abstract cosmic effect into a story form to help him understand and study the concept of planetary influence on his world. This way,

---

217 C.L. Barnhart, ed., *The American College Dictionary* (New York: Random House, Inc., 1962) 65.

the concept could be personalized and given life in our view of reality, and the evolved and unevolved expression of this energy or force could be attributed to the characters or archetypes in different stories that could help clarify the understanding of our relationship to both ourselves and the outer world.

In the most elementary terms, the natal birth chart is called a horoscope, Greek for "picture of the hour," for the time of birth of an individual and can represent the sky as seen from the date, time and location of a person's birth as well as the birth of a country, corporation, marriage, or any event for which information is needed. This "picture of the hour" is composed of planets, a Greek word for "messengers," which are energy vortex points that seem to have specialty functions as observed consistently for over 10,000 years. These planets are located in certain star constellations that comprise the "zodiac," Greek for "wheel of animals," because they have such names as Aries, the ram; Taurus, the bull; Cancer, the crab, etc. Actually, in modern times it is believed that the amusement park ride called the "merry-go-round" got its inspiration from the zodiac. Astrology is a blend of understanding the gravitation and magnetism of the earth's relationship to these heavenly bodies contained in our solar system and the Universe beyond. These forces are far greater than what meets the eye. They act like a symphony of invisible crosscurrents and alignments that can affect the earth and the physical, emotional, mental and spiritual evolution of its inhabitants.

As we look at astrology through contemporary eyes, it is really not so implausible to regard planets as functioning in a manner similar to man-made satellites. Just as there are many types of satellites with different specialties and purposes such as communications, weather forecasting, spying, etc., so it is with the metaphysical properties and specialties of planets. Planetary energies trigger many forms of awareness in the human species, both on a collective and an individual basis. The trick is to use the right satellite for the right intention or purpose.

The Birth of Venus and the Rose

The birth of Venus in mythology is very unusual and telling. Venus was born out of conflict, from the violence between father and son, Uranus and Saturn. The myth of her birth takes place when Uranus rejected his own creations in the form of his children by Gaia. As soon as each child, or creation, was born, he did not find them as perfect as his original concept of them, so he repeatedly shoved each child, as it was born, back into the womb of Gaia, Mother Earth. It is important to remember that this is esoteric language for an archetypal force springing out of chaos. It is deliberate in its attempt to represent that those we would call "gods" are not subject to the same rules of human procreation.

Weary of having her own children rejected by the 'creative force', represented by the Sky God Uranus, Gaia instructed their son Saturn, or Kronus in the original Greek, to await the approach of Uranus and castrate him. This is a deep allegory to demonstrate the consequences of thoughtless rejection of one's own creations, thus sacrificing the possibility of evolution to the pursuit of perfection.

The legend continues that Saturn threw the severed genitals of his father Uranus into the sea where a great and turbulent roiling of the waters began. The agitation of the waters increased to the point where a great foam was produced, and from this enormous foam, standing on a seashell, emerged the beautiful Aphrodite, which means "she who is born of the foam," or Venus, as we call her in Western astrology.[218] Among her many lovers were Mars, Hermes, and Adonis. In a jealous rage, Mars transformed himself into a savage wild boar and gored Adonis to death. Roses sprang forth from her tears at the death transformation of her lover or from the actual drops of his blood as they fell, depending on which version one chooses. The myth varies from region to region throughout the vast Greek Empire. Henceforth, the rose, the symbol of Venus, like the birth of Venus herself, became associated with beauty being born of violent transformation.[219]

Venus, Phi and the Rose

The rose has always been associated with Venus in Western astrological mythology. Just as the rose is a symbol

---

218 Ariel Guttman and Kenneth Johnson, *Mythic Astrology*. 41-51.
219 Ibid. 51.

of serene beauty, so also is the planetary influence of Venus and the signs that it rules, Taurus and Libra. Wherever we see Venus in the astrological chart, either where in the chart it was at birth or by transit, or where it is located in the chart on a later date, we also associate the symbol of the rose. It points to areas of our lives where we can embrace healing and diplomacy. It shows us that Beauty can be the doorway to higher consciousness.

We can also look at the rose as an analogy for the relationship between Body, Soul, and Spirit.[220] If we look at the beginning of the growth of a rose, we see a hard shell or sheath, representing the physical body, which is supported by a stem, or cord, attaching it to the earth. Eventually, the petals of the rose burst forth from the covering of the bud and open into their full magnificence, still nurtured by the vessel and support of the stem that connects it to its source. The rose exerts and exudes a magnetism that attracts other life, such as bees and humans, to assist in their evolution. Eventually, the petals dry and fall away and yield to the production of the seed, representing the spirit, from which the cycle can once again bring forth beauty into the physical world. This is a constructive manner in which to view Venus in the natal or transiting birth chart.[221]

The spiral of a seashell, the wave in the ocean, the shape of a galaxy, the needles of a pine cone and the petals of a rose are but a few of countless examples that demonstrate a number system discovered in the middle ages by an Italian

220 Lisa Tenzin-Dolma, *Understanding the Planetary Myths* (London: Quantum, 2005), 53.
221 Ibid. 56-57.

mathematician named Fibonacci, more properly called Leonardo of Pisa. He was born in Pisa around 1175 and died around 1250. A contemporary of St. Francis of Assisi and Marco Polo, he was educated by the Moors in Northern Africa during his youth when Europe was awakening to the Age of Exploration. He introduced the decimal system to Europe and was one of the first to introduce the Hindu-Arabic number system to European scholars, thus playing a significant role in causing Europe to change from the old and unwieldy system of Roman Numerals to Arabic numbers.[222] The interesting concept of the Fibonacci number system is that it is based on the numerical relationship of the spiral rather than the circle. The system is really quite easy to understand. A Fibonacci number is always the result of two preceding numbers that are in natural sequence with each other. The basic system is 1, 1, 2, 3, 5, 8, 13, 21, etc. Note that "2" is a combination of 1+1, the "3" is a result of 1+2, the "5" is a combination of the preceding 2+3, the "8" is a sum of 3+5, etc.

This formula is called "phi" and relates to the spiral. Another mathematical formula that we are all taught in school is that of "pi", which relates to a circle and its diameter. Now, this is important: the formula "phi" relates to the "spiral of life", which represents evolution and enlightenment; the "pi" formula relates to the circle, which is a loop or infinite manifestation of the status quo.

222 Richard A. Dunlop, *The Golden Ratio and Fibonacci Numbers* (Singapore: World Scientific Publishing Co. Pte. Ltd., 1998).

In terms of astrological psychology, defined as the relationship of the human subconscious to planetary movement, every time a planet returns to where it was when a person was born, there is an opportunity to take an evolutionary step upward in awareness and consciousness. If the person does not progress in awareness during each planetary return, he is symbolically "going around in circles". As Venus orbits the earth, it makes eight inferior conjunctions that occur when Venus is exactly between the Earth and the Sun every eight earth years, and if plotted on a graph, these conjunctions form a pentagram, a five pointed star, which is the symbol of the "Goddess." In terms of geometry, a pentagram is composed of five golden triangles. All of these triangles symbolically connect our perception of the material level of awareness with the cosmic or spiritual understanding of our relationship to the Universe.

Whenever Venus returns to the position in an astrological chart where it was at birth, a chart can be calculated for that moment to forecast the areas of life that will inter-relate with the current Venus cycle of the individual. There is a phi or Fibonacci relationship between the Earth and Venus in terms of their respective orbits. For example, the earth orbits the sun 8 times during the same time frame that Venus orbits the sun 13 times. 8 and 13 are phi, or spiral, evolutionary numbers.

Venus and Taurus

Venus rules the signs Taurus and Libra, and in astrological practice, the qualities attributed to a planet also apply to the zodiacal signs that it rules. Venus is inextricably associated with the rose, and since Venus rules the sign of Taurus, the rose is also associated with Taurus. The part of the body associated with Taurus is the throat and therefore the voice and the fifth chakra. People with prominent Venus, or with many planets in Taurus, or those whose rising sign is Taurus usually have very pleasant voices and speak in well-modulated phrases. There is a solidity to the speech and a very grounded style of communication.

Venus rules the second house of personal values in the natal chart. People with Venus or Taurus located in this area of the chart resonate with quality and durability. It is interesting that Venus, in astronomical terms, rotates on its axis from west to east instead of from east to west like most of the other planets. Therefore, if we were living on the planet Venus, the sun would rise in the west and set in the east. This reverse direction of axis rotation causes Venus to radiate strong magnetism. Those with prominent Venus energy in the chart usually draw things to them in terms of resources and helpful people. They collect and amass possessions and people and have great difficulty throwing things away.

The Rose, its Connotations in the East and the West

The rose can bear a similar description. It is a flower of quality and, while on the vine, longevity. Roses when picked for a bouquet usually require a more formal container than most other flowers. While each flower is perfection in itself, its quality can stand alone in a single bud vase and inspire a great appreciation and contemplation. Also, since people with planets in Taurus love abundance, they present roses, often in large bouquets by the dozens!

Being always mindful that Venus is associated with the throat chakra, we come across a more profound association with the potential of the rose. In Kabbalistic or Hebrew astrology, this fifth chakra is called "Da'ath" and is considered the doorway to the Underworld where both psychological and spiritual evolutionary transformation takes place. Hebrew astrology considers this Da'ath to be the energy point on the human body that serves as an initiation chamber to gain the wisdom to determine the difference between spiritual darkness and light, a process that can seem terrifying. The rose is connected both with romance and eros. Remember, the rose has thorns! They teach us that the rose must be approached with respect in order to fully hold and experience the gift of its beauty.

The symbolic meaning of the rose in Western civilization is equivalent to that assigned to the lotus blossom in ancient Egypt, India, and the Far East. It is the symbol of the elevated level of consciousness of the human who has awakened to his true identity. The fragrance of the rose has been long prized in many cultures, especially that of Persia and Bulgaria. These cultures are closely identified with essential rose attar or fragrant oils. To make one ounce of rose essence, over sixty pounds of rose petals are distilled. The rose is a prize and one of the hallmarks of idealized beauty.

Venus and Libra

The sign of Libra is the other domain or sign ruled by the planet Venus. While the aspect of Venus that we see in Taurus is sensual, acquisitive, and erotic, Venus in Libra represents the diplomat, the keeper of peace, the patron of fine arts. Remember that Venus represents beauty born of conflict as she rose out of the sea, standing on a seashell. Incidentally, the wave and the shell are two examples of spiral numerology beautifully immortalized in the famous painting by Botticelli. Here in Libra, the planet is concerned with the conflict of balancing relationships, whether between nations or individuals. It is no co-incidence that sending roses to promote romance and harmony is a long established tradition, especially on Valentine's Day.

Actually, Valentine's Day has its roots in very ancient Greco-Roman times when the celebration of the marriage of Juno and Zeus were honored in early February. It was a time of celebration of the goddess Venus and her son Eros and the magic of romance. Later, during the reign of the tyrannical emperor Claudius of Rome, the tradition was for young men and women to lead very separate lives from each other. But during the festival of Juno, the young maidens would put their names in a jar, and the young men would choose the name of a maiden who would then be his significant other for the following year. If they fell in love, they would marry; if not, there would be another year and another name to draw. Claudius outlawed this celebration, but a bishop, St. Valentine, would continue to secretly marry the couples against the emperor's wishes.   Claudius had St. Valentine beheaded on February 14th around the year 270. After the death of Claudius, the festival was reinstated, and during this time of the old pagan days of worship to Venus and Eros, the young men would once again choose a name from a jar, picking a young maiden to court for the following year. The maiden chosen was afterward known as his "Valentine."

The Rose and the Divine Feminine

Venus has long been revered as one aspect of the Divine Feminine in religious and spiritual practices from earliest times.   During the time when Christianity was beginning to pervade Europe, a representative of the Feminine was necessary to balance the severity of the masculine aspect of Christianity. The preceding cultures and religions were very oriented towards honoring the Goddess energy in their rituals and symbols. Hence, the increased veneration of the Virgin Mary and Mary Magdalene became more established in Christian theology of the time. The rose was not allowed to represent the Virgin Mary because of its long association with Venus, a Goddess of pagan times. Therefore, the lily was associated with the Virgin Mary because of its connotation with the concept of purity. The rose became associated with Mary Magdalene because of her relationship to sensuality, a basic quality associated with Venus. This is evidenced in the magnificent stained glass windows and cathedral sculpture throughout medieval Europe. Prime examples of this are the vividly colored "rose" windows in such major cathedrals as Notre Dame de Paris and Chartres. Many of those cathedrals were erected on the original sites of Goddess worship, which also served as ancient astrological observatories. There, particular attention was paid to the dual qualities of Venus: because of orbital synchronicities, sometimes Venus is seen as the "morning star" and sometimes as the "evening star."

Conclusion

Both in the language of sacred and cosmic geometry, the rose remains a symbol of beauty, combining the numerology of the macrocosmic and the microcosmic universe of the golden spiral of phi discovered by Fibonacci. Just as the meditative contemplation of the form, structure and substance of the rose can lead to a greater reflective and parallel understanding of ourselves, so by harmonizing with those planetary transits that eternally work their cosmic effect on the human condition, we might also open our understanding of unfolding growth. We only have to awaken to the lesson of the pattern of the petals of the rose, to follow the spiral of evolution instead of the circle of limitation.

Today, as human understanding of the art and science of astrology explodes with new insight, ancient myths become clearer to us. The balance necessary to maintain an evolving insight into creating our lives, rather than merely administrating life events into some understandable pattern, rests in understanding the sacred symbols of the distant past. The magic of today is the science of tomorrow.

The rose is prevalent in every astrological myth that involves understanding the vessel of feminine beauty. Perhaps it is best summed up by a little memorial plaque that graces a bench in New York's Central Park:

"Myths are stories that never happened, but always are."

7

Roses in Turkish Literature and Culture(1)

By Gamze Demirel (2)

Accepted as meritorious and holy, nature, mirroring traces of God, has been a source of inspiration to artists for centuries. The Turkish people revere nature and especially the flower known as the rose. Writing in 2004, Annemarie Schimmel, in a phenomenological approach to the study of Islam, saw the rose as the most precious of garden flowers in Islamic culture and the "sultan" of the genre. "Especially the red rose," she said, "indicates Divine Greatness" (39).

The rose has a remarkable place in the Islamic world of the imagery as narrated in the Qur'an, the Holy Scripture of Islam, and secondary Islamic narratives. The story of the Prophet Abraham, for instance, has given birth to the rose-fire-Abraham trio in classical Turkish poetry. Prophet Abraham, one of the Qur'an's greatest seers, also plays a significant role in the Christian Holy Bible, but his story begins when God orders him to leave his land for an unknown place that God will show him (Hebrews 11:8). The years of Abraham's youth are elaborated in the Qur'an in chapters 2:258–260,

6:75–84, 19:41–50, 21:51–70, 26:69–104, 29:16–25, 37:83–98, and 43:26–27. According to the Qur'anic account and further elaborations of the Prophet Muhammad, Abraham's father was an idol-maker for a self-acclaimed divine king. After Abraham demolished all the idols in the shrine, the king decided to punish him by throwing him into a huge fire. But instead of burning the victim, the flames turned into a rose garden.

In classical Turkish poetry, this image of Abraham, together with the Prophet Joseph's story, provides source material for numerous allegories of the rose.

Joseph, another prophet mentioned in the Qur'an, is the subject of a coherent narrative told in a chapter that bears his Arabic name, Yûsuf (*The Qur'an*, Chapter 12). The figure corresponds to the Biblical Joseph, the eleventh son of Jacob (Ya'qub). Believed to have been a very handsome and attractive boy, he is the smartest among his brothers, so clever in fact that jealousy caused them to throw him into a well. Rescued by a passing caravan, he was brought to Egypt to be sold as a slave to the King's finance minister (Potiphar in the Biblical account). When Joseph matured, Zuleikha (Potiphar's wife) fell in love with him and wanted to seduce him. His resistance lead her to tear his shirt from behind. In Classical Turkish poetry, this incident is likened to the blossoming of the rose; we see the emotional assault as a tearing off of petals.

Another Qur'anic and Biblical figure associated with roses in Classical Turkish poetry is the Prophet Solomon. In

Islam, it is believed that, God willing, the wind would provide this king's transportation. Needing to travel, Solomon would simply mount his throne, command the wind, and be carried even to the farthest distance in the realm. In Classical Turkish poetry, his throne is likened to rose petals flying in the air with the blowing wind. And the shape of the ring that he used to govern birds, jinns, and all creatures is likened to the rosebud. The blooming of the rose is also used as a metaphor for the resurrection of the dead when Israfel (in Hebrew Raphael) announced the coming of the end of the world by blowing the trumpet called Sûr the second time. It is believed that his first blow would bring the Apocalypse and his second start resurrection in the Hereafter.

In Turkish Islamic tradition, however, the rose represents most powerfully the Prophet Muhammad. Indeed, the Prophet Mohammed even sweats the scent of a rose; therefore, Muslim Turks hold, if you sniff a rose, you have done a good deed, an attitude that helps explain the intimate love of Turkish people for nature and especially for flowers. Many Turks believe that each flower has its own language to help explain a person's loves and sorrows. "One can discover leaves of dried flowers like roses and violas while thumbing through the pages of old books" (Onay, 2000: 221). These relics, possible gifts to express love and interest, had then been hidden away as souvenirs.

Not only blooms but also by-products serve as cultural coin. Traditionally, Turks put rose oils into flasks for traveling. Turks serve rose water to guests in inlaid cases called *gülabdan*, and they wash mosques with rose water. A person's blush is likened to the color of a rose, emphasizing that the rose, as well as decency and shame, has profound significance in Turkish culture.

The name of the rose and its derivatives have titled many books published in Turkey about Turkish culture. For example (*Gülbün* [Rosary], *Gülşen* [Bed of Roses], *Gülzar* [Rose Garden], *Gül-deste* [Bunch of Roses] etc.). The ceremonial cardigans Mawlawi and Bektashi (3) people used to wear for rituals were called "Bunch of Roses." Gravestones, too, were embellished by patterns of the "rose." The rose symbolism occurs many works of Turkish literature and art, often standing for the palace. As Kortantamer (1993) writes in her analysis of Old Turkish literature, "the 'rose' makes its whereabouts a center, s sort of palace, thanks to its shining like the sun" (413).

Because the rose blossoms in spring and brings beauty, color and happiness to gardens, another name of the spring in Turkish culture is "the season of the rose":

Nevbahâr oldu gelin azm-i gülistân edelim

Açalım gonce-i kalbi gül-i handân edelim

(Bâkî [4] - G [5]. 323, b. 1) (in Küçük, p. 299)

(In English: Now the spring has come. Let's go to the rose garden, open the buds of our hearts and create a smiling rose [that is, a rose in bloom].) (*Yesirgil, 1963: 57*)

Gül devri ayş eyyâmıdır zevk u safâ hengâmıdır

Âşıkların bayramıdır bu mevsim-i ferhunde-dem

(Nef 'î [6]- K [7]. 15, b. 3) (in Akkuş, p. 94)

(In English: Time of the Rose, spring in other words, is for feasting, gallivanting, living on the primrose path; this season, an era of happiness, is the holiday of lovers.)

Turkish families who give considerable thought to naming their children often call them "gül," that is, "rose" and names evoking roses (like *Gül, Gülcan, Gülben, Güllü, Gülsever, Goncagül, Sadegül, Esengül, Badegül, Birgül, Şengül, Songül* etc. (especially but not only girls). Similarly, many mosques, districts, streets and avenues have such names -- clear evidence of the importance granted to the rose in Turkish culture.

Moreover, numerous proverbs and idioms include the word "rose" in our language: *Dikensiz gül olmaz* (There is no rose without its thorn), (8) *El bebek gül bebek* (blue-eyed boy), (9) *gül gibi bakmak* (look after like a rose), (10) *gül gibi geçinmek* (get along well), (11) etc.

As a generic, "gül" in Turkish can mean all flowers, as seen in certain modes of Classical Turkish music: "for instance Gonca-i Rânâ, Güldeste, Gülizâr, Gülşen-i Vefâ, Gülzâr, Gülruh" (Ayvazoğlu, 1995: 100). Many classical compositions include the word "rose" and its derivatives. Rosebud (*gonca gül*) expressed in the following lines means lover, liver (*ciğer*) means heart and break in heart (*yâre*) means heartache:

"Bir gonca gülün yâresi vardır ciğerimde" (Dede Efendi) (12) (Literally, I have a broken rosebud in my liver.) The twentieth-century Turkish historian Reşat Ekrem Koçu (1905-1975) says in one of his soulful articles titled „Rose Oil" that the best ballads of Rumelia were sung while roses were boiling (in Ayvazoğlu, 1995: 198). Here is a ballad:

"Ellerinde gül oya/ Gülmedim doya doya

[In English: Rose lace in your hands/I have never laughed to my heart's content] (in Ayvazoğlu, 1995: 184).

Furthermore, the hymns and blessings chanted loudly that had been prevalent among Janissaries, Kalenderis (a group of Dervishs), Bektashis and Mevlevis were called *Gülbang. Gülgeşt* evokes wandering in rose gardens in those days. The circular shaped ornamental metal or wooden plate with a hole in the middle for a door knocker is also called *ağızlık gülü* (nozzle rose) (*Encyclopedia of Islam*, 1996: "The Rose").

The rose, gesturing toward the other realm (that invisible, unknown, secret world) by means of its color, aroma and shape, has also been used to explain mortality in earthly life. Life is short, almost too short for growing roses:

Bir zebân-ı hâldür her yaprağı fehm itseler

Perde-dâr-ı hâk olanlardan virür ahbâr gül

(Fuzûlî (13) - K. 9, b. 33) (in Akyüz, et al., p. 46)

(In English: Each living rose petal suggests others buried under the soil -- the dead -- each having lived and died in its own distinct way.)

As to the rose's expressive utility, in olden times when directly revealing such wishes to parents would have been deemed disrespectful, Turkish youth announced their desire to marry by using various symbols. For instance, a girl or boy donned "socks with roses" to tell the family he or she had reached marriageable age. Similarly, it is common tradition to swaddle babies with rose-patterned clothes. "It is also known that rose petals were scattered on beds in the past" (Onay, 2000: 221). In addition, roses figure significantly in Turkish handicrafts, rose-shaped patterns (rose lace), for instance, or a heavy, silky type of fabric called "*Güllü Dîbâ*" imprinted with the flower (*Encyclopedia of Islam*, 1996: "The Rose").

Furthermore, throughout Turkey rose rhymes and references are used in the names of protagonists of famous anonymous love poems such as *Varka ile Gülşah* (Varka and Gülşah), and *Hüsrev ile Gülruh* (Hüsrev and Gülruh), which were written in the 14th century masnawi style (consisting of an indefinite number of couplets, with the rhyme scheme aa/bb/cc). As in many other cultures, "6th of May is accepted as *Hıdırellez* (celebration of spring) in Turkish culture. In *Hıdırellez*, people leave notes containing their desires under rose trees where prophets Hızır (an immortal person thought to come in time of need) and İlyas (Elijah) are believed to have met and throw these notes into the sea, to the kingdom of Hızır, while the sun is rising" (Gezgin, 2010: 82). Another source states that "... the oil roses blossom in May and are boiled in the boilers laid on the 6th of May, namely *Hıdırellez*, till the end of May" (Ayvazoğlu, 1995: 198).

Rose oil and rose water were not merely aromatic and therefore beautiful but also useful: they belonged to the inventory of physicians in the Ottoman period beginning at the end of the 13th century. Sources confirm that roses in the gardens of Sufi monasteries provided medical benefits:

Bulunur her derde istersen gülistânda devâ

Hokkasında goncenün san kim şifâ cüllabı var

(Fuzûlî- G. 74, b. 4) (in Akyüz, et al., p. 167)

(In English: If you wish, you can find a cure for all your problems in the rose garden, in curative rose water in the pot of the bud.)

If diarrhea is serious, apply to the stomach a poultice made of the pulp of rose, celery, anise and cumin seeds; or of cumin seeds and rose seeds in vinegar; or oat seeds boiled in vinegar (Kahya, 1995: 236).

Sometimes, children's digestion weakens. This illness is cured by applying rose water, lily and myrtle (ibid. 240). For instance, when saffron is mixed with rose, camphor and coral, it penetrates into the heart (ibid. 38).

Should these mixtures fail and health fade, the rose at least goes on: you find in Anatolia innumerable tombs named *Gül Baba* (Father Rose) expressing the flower's symbolic meaning in Turkish Sufi thinking. The rosebud refers to unity (union, God) and the rose in bloom to abundance (wealth, everybody and everything on earth, as well as all that prevents human beings from reaching their creator).

Jam and syrup are made from roses. Desserts like *güllaç* (starch wafers filled with walnuts soaked in milk and rose water), *su muhallebisi* (starch pudding with rose water), *güllabiye* (halva with almond and rose water and garnished with rose petals) were the indispensable flavors of Ottoman cuisine.

The rose has had an important place in the daily lives of Turkish people from the Ottoman period till today. Commonly used in cosmetics (perfume, soap, etc.) the flower plays an important role in generating Turkey's agricultural income as well.

The rose – a generic for all flowers -- is also a common symbol of love as well as a sign of mystery and sanctity. The legendary love between the rose and the nightingale is inexhaustible in Turkish writing, as in world literature in general. Its color, smell, shape, thorns, brief lifespan and suggestive buds have made it a recurring trope in classical poetry. The classical Turkish poem best known in the West is the masnawi by 17th-century Ottoman poet Kara Fazli called *Gül ü Bülbül* (The Rose and The Nightingale). Many Ottoman poets wrote odes with the "rose" rhyme including Necâtî (15th century) and Nev'î (16th century). But among these odes, one of the best is Fuzûlî's written for Suleiman the Magnificent (1495-1566), the tenth Ottoman Sultan, under the penname Muhibbî. The stanzas' rich image world offers a different play on words in nearly every use of the term "rose" in sixty-two couplets (*Encyclopedia of Islam*, 1996: "The Rose"). And in the following couplet, the 18th century-Ottoman poet Nedim loves the rose as a form of addressing his beloved:

Gülüm şöyle gülüm böyle demektir yâre mu'tâdım

Seni ey gül sever cânım ki cânâna hitâbımsın

(Nedim- K. 19, b. 6) (in Macit, p. 74)

(In English: It is my habit to call the beloved "my rose," so I love the rose for giving me a name for my beloved.)

In spite of cruelty, arrogance or disloyalty, the lover welcomes the beloved as a rose, an attitude also reflected in the tradition of fixing a rose on the head or turban (*Encyclopedia of Islam*, 1996: "The Rose"):

Öyle pinhân eylemiş gögsinde sırr-ı ışkunı

Kim ayağından asarlar eylemez izhâr gül

(Fuzûlî- K.9, b. 24) (in Akyüz, et al., p. 45)

(In English: The rose keeps the secret of your love in its heart and cannot give it away for fear of being hanged by the foot, that is, being pinned upside-down on the turban or chest.)

Furthermore, lovers' tears are rose-colored (red) or in sprinkled droplets of rose water:

Urunca şâne gîsû-yı hayâl-i yâra müjgânım

Gülâb-efşân olur yâd-i ruhıyla çeşm-i giryânım

(Nef'î- Müfred [14] s 1) (in Akkuş, p. 352)

(In English: Envisioning my beloved, my crying eyes sprinkle rose water when I recall her face.)

The lover can't take his eyes off the cheek of the lover that resembles a rose:

"Günde yüzün elli kerre görsem de,

Tamahkârım, gözüm doymaz gül yüze"

(in Özalp, p. 468)

(In English: Even seeing the face of the beloved fifty times a day wouldn't satisfy me, for I am greedy. I cannot be happy without those rosy cheeks. In other words, the lover wants to remain always in the presence of his beloved.)

We might also say that this lover is enflamed, associating roses and fires, a plausible interpretation since classical Turkish literature often reminds us that whatever stands upright overcoming obstacles is a vertical element, and these vertical elements are celestial and divine. Both roses and fires partake of this association.

The similarity of color between roses and fires has been the source of many images in Turkish poetry:

Gül âteş gülbün âteş cûy bâr âteş

Semender-tıynetân-ı aşka bestir lâlezâr âteş

(Şeyh Galip [15]-G. 139, b. 1) (in Kalkışım, p. 327)

(In English: The rose is fire, the rose sapling is fire, the rose garden is fire, and the river is fire. Fire is enough for the salamander-like lovers in a tulip garden/ The rose, the rose sapling, the river turning red in the evening and the poppy garden share their color with the blaze.)

The rose pattern commonly appeared in post-Tanzimat (16) Turkish literature, written in a period known for accelerated Westernization in all aspects of the declining Ottoman Empire. Thus, despite considerable change and novelty, the rose remained an indispensable literary trope. Recaizade Mahmut Ekrem (1847-1914) expresses his love for the rose and the nightingale in his book *"Pejmürde* (Dishevel),"* while nationalist poet and writer Namık Kemal (1840-1888) uses the rose as the symbol of nation and flag in his poem *"Vaveyla":*

Âh böyle gezer mi hiç canan?

Gül değil arkasında kanlı kefen...

Sen misin, sen misin garîb vatan?

(Namık Kemal, Vaveyla, Nefta 1)

(http://www.antoloji.com/siir/siir/gunun_siiri.
asp?year=2011&month=7&day=15)

(Alas, does the beloved ever go around like that?

Behind her is not a rose, but a bloody shroud

Is that you, is that you? Oh lonely homeland!)

The rose maintains its irrevocable fame in Turkish literature in the Republican Era that started with the establishment of the Republic of Turkey as a successor to the fallen Ottoman Empire in 1923. The Second New (*İkinci Yeni*) current in Turkish poetry that was shaped in the 1950s by poets like Edip Cansever, İlhan Berk, Cemal Süreya, Sezai Karakoç and Ege Ayhan frequently used rose metaphors to bring esoteric language back to Turkish poetry which had been de-popularized by the "Garip" literary trend, a movement to nationalize the Turkish language once the modern nation-state had taken root.

Therefore, Second New current poets like Behçet Necatigil (1916-1979) and Ahmet Muhip Dranas (1909-1980) challenged conventional patterns in language, forcing the limits of word ordering by changing or disrupting idioms. Emphasizing imagination and feelings in poetry, they privileged abstract metaphors and made the rose pattern an essential poetic tool.

Ahmet Haşim (1884-1933) was among the pioneers of symbolism in Turkish literature, even before the emergence of the Second New current in the 1950s. The rose is the flower he loves most, especially the image of "the bleeding rose" in "Merdiven" (The Stairs), a remarkable poem:

"Eğilmiş arza, kanar, muttasıl kanar güller;

Durur alev gibi dallarda kanlı bülbüller,

Sular mı yandı? Neden tunca benziyor mermer?"

(Ahmet Haşim/ Merdiven)

(Bent towards the ground, the roses are constantly bleeding;

The bloody nightingales stand on fire-like branches,

Have the waters been burnt? Why does the marble look like bronze?)

Another influential poet and writer of the early Republican era was Yahya Kemal Beyatlı (1884-1958), who mentioned the rose on various occasions in his poems, especially in *"Rindlerin Ölümü"* (Death of the Dervish) and *"Rindlerin Akşamı"* (Night of the Dervish), two of the most important works presenting the "rose" pattern in an impressive way.

Similarly, in *"Endülüs'te Raks"* (Dance in Andalusia) the rose gives its color to the entire piece:

The bell, the shawl, and the rose

All the speed of dance is in this garden

Andalusia is three times red on this night of pleasure

. . . .

Rose-skinned, ember-lipped, coal-eyed blackened with kohl

Satan says one should hug and kiss her a hundred times,

To the splendiferous shawl, to the enchanting rose,

From all bosoms to the bells filling all hearts

Zil, şal ve gül

Bu bahçede raksın bütün hızı...

Şevk akşamında Endülüs üç def'a kırmızı...

....

Gül tenli, kor dudaklı, kömür gözlü, sürmeli...

Şeytan diyor ki sarmalı, yüz kerre öpmeli..

Gözler kamaştıran şala, meftûn eden güle,

Her kalbi dolduran zile, her sineden

Rindlerin Ölümü

Hafız'ın kabri olan bahçede bir gül varmış;

Yeniden her gün açarmış kanayan rengiyle.

Gece; bülbül ağaran vakte kadar ağlarmış

Eski Şiraz'ı hayal ettiren ahengiyle.

Ölüm asude bahar ülkesidir bir rinde;

Gönlü her yerde buhurdan gibi yıllarca tüter.

Ve serin serviler altında kalan kabrinde

Her seher bir gül açar; her gece bir bülbül

öter

(Yahya Kemal Beyatlı, in Ayvazoğlu, p. 238in Ayvazoğlu, p. 238)

Death of the Dervishes

There was a rose in the garden where Hafiz was
buried

Each day it bloomed again with its bleeding color

At night, the nightingale wept till dawn

With its melody reminiscent of Ancient Shiraz

Death is a peaceful land of spring for the dervish

His heart fumes like a censer for years

And in his grave under cool cypresses

A rose blooms each morning; a nightingale sings

each night.

Rindlerin Akşamı

Dönülmez akşamın ufkundayız, vakit çok geç

Bu son fasıldır ey ömrüm nasıl geçersen geç

Cihana bir daha gelmek hayal edilse bile

Avunmak istemeyiz öyle bir teselliyle

Geniş kanatları boşlukta simsiyah açılan

Ve arkasında güneş doğmayan büyük kapıdan

Geçince başlayacak bitmeyen sükunlu gece

Guruba karşı bu son bahçelerde, keyfince

Ya şevk içinde harab ol, ya aşk içinde gönül

Ya lale açmalıdır göğsümüzde yahud gül

(Yahya Kemal Beyatlı,

http://www.siirperisi.net/siir.asp?siir=716)

Evening of the Dervishes

We are at the end of a no-return path (evening),
too late it is

This is the last episode, oh my life,

may you pass however you like it

Even if coming back to the world once again is
dreamed of

We do not want to be consoled with such a solace

When you pass through the grand gate

with its widely opening wings in dark space

And where no sun rises behind it

Then will start the night with unending silence.

Against the sunset in the garden of life, as you
like

My heart, be ruined either in desire or love

In your bosom, should bloom either tulip or rose.

Thus, the rose remains the beloved flower of Turkish literature with all its connotations. Sezai Karakoç (1933-), a contemporary Turkish poet, thinker, and politician, heralds the coming of a civilization shaped by Prophet Muhammad associated with the rose in *Gül Muştusu* (*The Gospel of the Rose*) (1974). The rose symbol is also the main topic of *Güller Kitabı* (*The Book of Roses* republished in 1992). This important work, written originally by the Turkish historian Beşir Ayvazoğlu (1953-) traces the role of flowers, especially the rose, in Turkish culture. He analyzes the feelings of Turks toward flowers and nature in general from nomadic tribes to great civilizations. Only in Nazım Hikmet Ran (1902-1963), a famous Turkish poet from recent times, do these feelings towards the rose seem negative in contrast to the tradition that had come before:

"Güle, bülbüle, ruha, mehtaba, falan filan/ karnımız tok/ Ve şimdilik/ gönül işlerine vermiyoruz metelik."

(English: We don't believe in roses, nightingales, spirits, or moonlight /and for now/we have no regard for affairs of the heart)

In sum, the rose has always attracted great attention as an important element not only in Turkish but also in world culture and literature as confirmed by innumerable legends and beliefs in both East and West. Turks in particular have integrated „roses" into every aspect of their lives. Indeed, Turkish culture has been shaped by the love of the "rose" in all its beauty and meaning. With its rich accretion of symbols extending from Adam and Eve and the Prophet Abraham to the Prophet Mohammed, the rose has a cultural background as old as the history of humanity. We might even say that, in literature, architecture, poetry, music, Sufism, calligraphy, illumination, engraving, house decorations, gravestones, ceramics, mural paintings, clothes, names, jewelry, behaviors, value judgments, thought systems and lifestyles, the rose is emblematic of the Turkish nation itself.

NOTES

1 I would like to give my special thanks to Semiha Topal for help translating this article into English.

2 Assistant Professor of Classical Turkish Literature in Süleyman Şah University, Faculty of Humanities and Social Sciences, Department of Turkish Language and Literature, Istanbul, Turkey.

3 Mawlawi (Mevlevi in Turkish); is a Sufi order founded in 1273 in Konya, Turkey, by followers of the great Sufi Mawlana Jalal Ad-Din Al-Rumî, known commonly as Rumi in the West. They are known as the Whirling Dervishes due to the special zhikr ritual named Sema. The Bektashi order was also founded in Turkey within the same time period (13th century) by the followers of Hadji Baktash Wali (commonly known as Bektaşi in Turkish).

4 Bakî was the pen name of the Ottoman Turkish poet Mahmud Abdülbâkî (1526-1600 AD).

5 "G" is the abbreviation of "gazel (ode)," one type of poem in Classical Turkish Literature and „b" is the abbreviation of „beyit (couplet)," one of the poetry units that combines two lines.

6 Nef'î (1572-1635) was an Ottoman poet and satirist most famous for his odes (kaside).

7 "K" abbreviates "kaside (ode)," one type of poem in Classical Turkish Literature.

8 Every good and beautiful thing has a disruptive aspect (Proverb).

9 Spoilt and sassy (Idiom).

10 1) Making a living without financial difficulty; 2) Looking after oneself in a good and clean way (Idiom)

11 1) Getting along really well and keeping in good with someone 2) Living comfortably without vast opportunities.

12 Dede Efendi (1778-1846): His full name is Hammamizade İsmail Dede Efendi, a well-known composer of Turkish Classical Music.

13 Fuzûlî (c. 1483-1556) was the pen name of the Ottoman poet, thinker, and writer Muhammad Bin Suleyman. He wrote odes (kaside) in Azeri, Persian, and Arabic.

14 Müfred is one type of ClassicalTurkish poetry.

15 Şeyh Galip (1757-1799): A Classical Turkish poet who was also a Sufi.

16 The Tanzimat, meaning reorganization of the Ottoman State, was a period of reformation that began in 1839 with the declaration of the imperial edict aiming to modernize the Ottoman State, and to integrate non-Muslims and non-Turks more thoroughly into Ottoman society by enhancing their civil liberties. The period ended with the First Constitutional Era in 1876.

REFERENCES

Akkuş, Metin, (1993), *Nef'î Dîvânı* (Collected Poems), Ankara: Akçağ Publishing.

Akyüz, Kenan, et al., (2000), *Fuzûlî Dîvânı* (Collected Poems), Ankara: Akçağ Publishing.

Ayvazoğlu, Beşir, (1995), *Güller Kitabı* (Book of Roses), Istanbul: Ötüken Publishing.

Gezgin, Deniz, (2010), *Bitki Mitosları* (Herb Mythos), Istanbul: Sel Publishing.

İbn-i Sina, (1995), *El-Kânûn Fi't-Tıbb,* trans. by Esin Kahya, Ankara: AKM.

Kurnaz, Cemal, *İslam Ansiklopedisi* (Encyclopedia of Islam), (1996), "Gül" (The Rose), İstanbul: T.D.V. Publishing.

Kalkışım, Muhsin, (1994), *Şeyh Galib Dîvânı* (Collected Poems), Ankara: Akçağ Publishing.

Kaplan, Mahmut, (1996), *Neşâtî Dîvânı* (Collected Poems), İzmir: Akademi Bookhouse.

Kortantamer, Tunca, (1993), *Eski Türk Edebiyatı: Makaleler* (Old Turkish Literature: Articles), Ankara: Akçağ Publishing.

Küçük, Sabahattin, (2002), *Bâkî ve Dîvânı' ndan Seçmeler* (Bâkî and Selections from His Collected Poems), Ankara: Ministry of Culture.

Macit, Muhsin, (1997), *Nedîm Dîvânı* (Collected Poems), Ankara: Akçağ Publishing.

Mazıoğlu, Hasibe, (1986), *Fuzûlî ve Türkçe Dîvânı'ndan Seçmeler* (Fuzûlî and Selections from His Collected Turkish Poems), Ministry of Culture a Thousand Essential Works, Ankara: Ministry of Culture.

Karahan, Abdulkadir, (1985), *Nef' î Dîvânı' ndan Seçmeler* (Selections from Nef'î's Collected Poems), Ankara: Ministry of Culture.

Nevzat Yesirgil, (1963), *Bâkî: Hayatı, Sanatı, Şiirleri* (Bâkî: His Life, His Art, His Poems), İstanbul: Varlık Publishing.

Onay, Ahmet Talât, (2000), *Eski Türk Edebiyatında Mazmunlar ve İzahı* (Imageries and Their Interpretation in the Old Turkish Literature), Ankara: Akçağ Publishing.

Özalp, Nazmi, (2000), *Türk Musikisi Tarihi-II* (The History of Classical Turkish Music –II), Ankara: Ministry of National Education.

Schimmel, Annemarie, (2004), *Tanrının Yeryüzündeki İşaretleri: İslama Görüngübilimsel Yaklaşım* (Deciphering the **Signs of God**: A Phenomenological Approach to Islam), trans. by Ekrem Demirli, İstanbul: Kabalcı.

8

"Dying, Laughing"
The Rose From Yeats to Rumi

By Frankie Hutton

We lovers laugh to hear "This should be more that and that should be more this" coming from people sitting in a wagon tilted in a ditch. Going in search of the heart, I found a huge rose under my feet, and roses under all our feet...

Rumi

(Translation by Coleman Barks)

This essay surveys timeless, metaspiritual qualities of the rose as reflected in classical poetry and literature. Serving as didactic guideposts for human beings, the renderings considered in this essay are no stranger to sophisticated academic communities. Yet and still the higher underlying messages are not so obvious. The passages to be considered in this essay transcend vast time periods and places to teach esoteric knowledge to those seekers who are awake and aware. Thus, from important distinctly singular writers of diverse cultures and epochs we are presented with the same messages in different format and creative style. A bevy of metaphysically astute writers have used the rose as a potent symbol to fashion poems and short stories that provide insight to higher consciousness. The writers whose work is surveyed in this essay apparently knew full well that the rose is an exceptional symbol "planted" on earth to teach all who are ready about real life, death, and higher spiritual realms. Thus the rose is unlike human guides who appear at strategic times when one is on the path; it is ever present, ever ready, ever useful and ever beautiful for those who are wise enough to pay attention to it.

Intriguingly, because no one knows for sure exactly when the rose first appeared on earth, the lovely flower has come to have the staying power and clout of such enduring symbols as the cross, the oceans, the sun, mountains and so on. Eons ago, although we cannot be sure precisely when, the quintessential rose took its place beside entities that far overstep the material mind toward the ineffable and everlasting. Explored in this essay as a profound metaphor in timeless literature and poetry, the rose easily, pervasively transcends what is known of its potent botanical qualities deep into the mysteries of the universe and human existence. For it is in treading timeless literature connected with this magnificent flower that it can be revealed in full splendor. Careful review and reflection of rose-embracing literature and poetry reveals that the flower in all its perfection is a direct link to the Source of All and Everything. For those who have eyes to see, it comes into focus as one of the most long lasting, widely used and beautifully powerful symbols of all time.

Known for its beauty, medicinal qualities and as a provocative organization, club,[223] and municipal[224] symbol for eons, the rose has also been a powerful metaphor in movies,[225] in dance[226] and in works of fiction and nonfiction.[227] Barbara Seward's study of the rose in British literature is used pivotally here to reinforce the notion of the flower's wide use in popular literature. Allegorically and metaphysically, some of the works used in Seward's literary analysis are selections that transcend the material world. Surely she was aware of this. In *The Symbolic Rose* Seward uses Dante Alighiere's *Divine Comedy* as the primal literary antecedent to engineer a skillful analysis of the rose in selected works of British authors William Butler Yeats (1865-1939), T. S. Eliot (1888-

223 For example, the Girlfriends, Inc. has used the rose as a symbol since the organization was founded in 1928.

224 The Island of Rhodes in Greece used the rose as a symbol on coins in 1200 B.C.

225 For instance, the movie "V is for Vendetta" in which rose petals are used as a powerful symbol of good overcoming evil. Starring Natalie Portman and Hugo Weaving and directed by James McTeigue, "V" was released in 2006.

226 In Nandor Fodor's, *Between Two Worlds_* (Parker Publishing Company, West Nyack, New York) see Chapter Two "The Riddle of Vaslav Nijinsky" that describes Nijinsky's superior dance skills and levitation in the ballet "Secret Rose." On page 22, it is described how "Nijinsky possessed the ability to remain in the air at the highest point of elevation before descending." Although Nijinsky suffered from mental illness later in his life, he is known to have had unusual talent connected to dance feats.

227 See for instance, see Umberto Echo's *The Name of the Rose*, Hartcourt Brace & Company (New York) 1980, a well-researched work of fiction and also the movie based on the this title starring Sean Connery and Christian Slater.

1965) and James Joyce (1882-1941), all masters of prose and poetry. Seward goes further to make the point that before Dante's *Divine Comedy* "no earlier writer had attempted to express through a single rose the diversity of meanings associated with" the flower.[228]   Likewise, Seward proffers that no one "had demonstrated the poetic genius" or fully reconciled "such seemingly disparate qualities as carnal and spiritual love." But Seward did not reach back far enough into the vast history and cultures of the globe to take note that the rose has been used in mystically fired literature long before William Butler Yeats and even long before Dante's *Divine Comedy*. What's more, it is apparent that Yeats was aware of the rose's connection to the Divine for he often wrote about both as corollaries.

In the *Symbolic Rose*, Seward, herself a British author, introduces Yeats's work from a nineteenth century foundation of theosophy and secret societies while Eliot and Joyce "acquire a good part of their meaning from associations with the rose of Dante..." – that is to say from Italy of the 14th century. Apparently all three author's works were influenced by Dante Alighiere's *Divine Comedy*_simply because it is a masterpiece.   The story is a classic in world literature and has been used as a teaching tool in the academy for hundreds of years, but mastering *The Divine Comedy* is no easy feat and there is no attempt to accomplish its complete analysis here except to make clear that Dante's journey points to what

228 Barbara Seward, *The Symbolic Rose*. (Dallas, TX: Spring Publications, 1954). 37.

Helen Luke has described as toward the "vision of God" and the single truth that "without divinity there can be no conscious humanity and without humanity the divine remains an abstraction."[229] This point will be underscored in various ways in this and in the two essays following. A short recap of Alighiere and his *Divine Comedy* is useful here to show the epic poem's connection to the symbol of the rose. Under the influence of the Roman Church, Dante Alighiere's inner struggle was profound as was his exile from Florence between 1307 and 1321, the time during which *Divine Comedy*_was written. The exile, simply put, was the result of bad political and church politics of the time. The ruling political party, the Guelphs, was supported by Dante's family, but split over the nature and power of the papacy. This divide resulted in a lasting feud in which Dante, who is said to be one of the world's oldest poets, protested papal policies. Battles between the two groups went on for years and Alighieri went into extended exile.

Dante Alighieri's sets his epic saga to begin during the Easter Season on Good Friday in the year 1300 and tells of a man on a journey in which he encounters characters and symbols related to salvation and damnation. Significantly, the entire work is based on the number 3, an important number in metaphysics that suggests human beings must rise above the dualities confronting them in this world to a higher place that is sans tension, fighting and foolishness. Thus, in sacred

229 Helen M. Luke, *Dark Wood to White Rose: A Study of Meanings in Dante's Divine Comedy*. (Pecos, NM: Dove Publications, 1975). 11.

geometry when a third force is introduced, two initial forces are neutralized. On his journey Dante is "turned back by three beasts—the leopard, the lion and the wolf—by his love of pleasure, by his fierce pride, and by the terrifying latent greed and avarice of the ego."[230] These are "beasts" that most of us have to conquer on the material realm, in what we refer to as "life," in order to move up and on to the Divine. Verses in <u>Divine Comedy</u> are arranged in units of three lines and the epic is composed of three books, *Inferno, Purgatory* and *Paradise,* all phases Dante encounters on his spiritual journey.[231] Stiff moral judgments are forced on Dante as he reaches Earthly Paradise. In paradise, Dante encounters Beatrice, the sacred feminine, his guide and inspiration. It is she who guides him toward God, revealed in "an ineffable vision of light."[232]   Of course paradise is a beautiful place where the atmosphere is in "perfect society moving in the harmony of the dance or forming the flawless pattern of the *rose.*"[233] It's easy to follow Dante's symbolism here, to imagine the rose as heavenly perfection.

Dante waited until near the end of the epic to introduce the symbolism of the rose probably for that reason, "heaven" does symbolize perfection for which we must strive as we work toward conquering the "beasts" of the material realm. Overcoming the dualities of the earth to the point that a third

---

230 Luke. *Dark Wood to White Rose.* 11.
231 Dante Alighieri's *Divine Comedy,* Translated by H. F. Cary in *Classics Appreciation Society Condensations* (np: Grolier, 1955). 451.
232 Ibid. 444.
233 Ibid.

force is introduced is not work for the weak at heart; it's daily work that must be done in earnest all the time, twenty-four-seven to use contemporary vernacular. And, depending on the progress made this time around, perhaps the striving continues in other incarnations as well. In Canto XXXII, Dante spells out, in seven steps or gradations the struggles one must endure to reach paradise, leaf by leaf as "throned on the rose." In this last Canto, Dante is really on to something profound that has been somewhat veiled from the masses for thousands of years. Actually, it hasn't been hidden too deeply for as we will take note later in chapter nine the knowledge is replete in sacred books and works globally, the most prominent of which is the Christian Holy Bible and in a variety of literature, including significant works of fiction. To stay on point, the pathway is mapped for those who work to undertake the journey through consistent work that revolves around unconditional love, service and the untiring desire to know themselves and truth.

As a symbol, the rose continues to be a marker on the pathway. The initiated know well as do those with long traditions of meditation and study in metaphysics have some sense that the phrase "leaf by leaf" alludes to the consecrated work that must be done individually, unremittingly to move up the ladder, through the full number of rounds to connect fully and everlastingly with what Christianity has labeled the Holy Spirit. This process has been laid out and even described in sacred literature such as in the Christian *Holy Bible* to be considered in the last chapter of this anthology, but it is also deeply hermetic knowledge that is easy to miss for the spiritually uninspired. Dante was inspired and wanted very much to be illumined. For as both writer and protagonist of the *Divine Comedy*, Dante Alighiere's epic underscores the pain and the beasts of life that we all encounter. His earthly or material body died in Revenna in 1321 while away from his beloved Florence. If Dante didn't make the full journey to eternal life or endlessness, his poem shows us that he was intimately acquainted with the path and knew of the impediments to reaching the Light. His epic has been and continues to be a staple in world literature[234], religious studies and philosophy. Peter Bondanella's analysis has shown that Dante Alighiere "was under the influence of Saint Jerome's Latin Vulgate version of the Holy Bible from which he draws 600 references to the epic poem. Alighiere, according to Bondanella's view, confronts the seven deadly sins of lust, gluttony, avarice, sloth, wrath, envy and pride in purgatory.[235] Although the work of the three authors Seward considered in her analysis speaks volumes about love, life, eternity, resolution and ultimately connection to the Source, it is Yeats's metaphysical connections that tip us off that his objective is much higher, to teach us something about full initiation, interestingly in the tradition of the rose as the ultimate metaphor, just as expressed by Dante Alighiere. Relatively, Yeats uses the rose a lot in his poems and short stories; it

234 Dante Alighieri, *The Inferno*, Translated By Henry Wadsworth Longfellow, Introduction and Notes By Peter Bondanella, (New York, Barnes and Noble Classics, 2003), p. xxxi.
235 Ibid., xxxv.

appears in "To the Rose Upon the Rood of Time," "The Rose of the World," "The Rose of Peace." This mission is revealed quietly in allegory, metaphor and symbol throughout much of Yeats's work. The question is why? Well, it is apparent from all we know about Yeats at this remove that he grew tremendously in understanding of mysticism. At times he seems to want his readers to know this and just as important, to tip them off that growth is possible for them too.

While Yeats's use of the rose appears to be the most esoteric of the three authors Seward explored, it should be noted tangentially that Seward also does a fine analysis of Eliot and Joyce's use of the flower. Where the work of T. S. Eliot is concerned, she shows that the rose is the "obvious symbol of whole and enduring resolution" even as it combines "the romanticism of a yearning, nostalgic, insatiable age with absolute, authoritarian standards of medieval times..."[236] Known for social criticism in both his essays and poetry, Eliot was born in St. Louis, Missouri (1888) and educated at Harvard University. He moved to England after taking his master's degree in 1910, thus he is sometimes claimed as a favorite son by both the United States and England; obviously Seward includes him in the latter geographic lot. James Joyce was educated at Jesuit schools in Dublin and after 1902 moved to England. He also lived in Zurich during World War I; but the extremes of the two—that is a solid catholic education pitted against the backdrop of the horrors

---

236
Barbara Seward, *The Symbolic Rose*, p. 156.

of a world war must have been very difficult for his psyche. Seward notes the conflict in his writing. The whole point is that all three great writers Yeats, Eliot and Joyce made full use of the rose in their prose and verse.

Clearly Yeats' use of the rose in one particular short story is the centerpiece of this essay. His prose treads higher spiritual realms in a timeless but brief story that is easy to recount here and easily instructive. Eruditely reasoned, Seward's analysis says that Yeats's work is a marker in the literary history of the rose. For it is Yeats Seward proffers who makes the first successful attempt since Dante's *Divine Comedy* to express tradition as well as personal meanings in the single symbol of the rose. Seward notes that Yeats used a "medley of artistic and religious concepts."[237] It is apparent that Seward was aware when she wrote *The Symbolic Rose* that Yeats had joined the Dublin Theosophical Society in 1886 and followed this association with the hermetic tradition of the Order of the Golden Dawn in London, then a highly intellectual secret fraternity.[238] The influence of the Theosophical Society on Yeats was probably as profound as this initiation into the Hermetic Order of the Golden Dawn, an offshoot of the Rosicrucians, another secret order established perhaps two centuries earlier. Both organizations must have played a role in a spiritual breakthrough for Yeats. The influence of these groups on his writing is undeniable. Although we cannot be sure exactly when Yeats started to be

237 Ibid. p.89.
238 Ibid. p. 89.

fully influenced by either one of the groups, it is noteworthy that the Theosophical Society was founded in 1875 by H. S. Olcott[239] and H. P. Blavatsky. In 1890, the year before she died, Blavatsky is said to have started a closed esoteric section of the Theosophical Society; Yeats was a member of this group too.

Olcott's address "My Stand for the Theosophical Society," a classic in Theosophical circles, puts the overall mission of the group squarely: "if I understand the spirit of this Society, it consecrates itself to the intrepid and conscientious study of truth, and binds itself, individually as collectively, to suffer nothing to stand in the way." The Society is a worldwide organization devoted to the study of ageless wisdom and encompasses the study of various religious traditions as well as philosophy, scientific theories and systematic spiritual practice. The three objectives of the Society are paramount to its hundreds of lodges and study groups that meet regularly: "To form a nucleus of the universal brotherhood of humanity, without distinction of race, creed, sex, caste, or color. To encourage the comparative study of religion, philosophy, and science. To investigate unexplained laws of nature and the powers latent in humanity."[240]

Thus we witness in Yeats's writing a real aim to get at truths, to reveal them to his readers, but in somewhat veiled symbolism. There is no surprise that he did this for Yeats

239 Richard Cavendish, *A History of Magic* (NY: Arkana, Division of Penguin Books, 1990) p. 142.
240 *Introducing the Theosophical Society*, Wheaton, Illinois. http://www.theosophical.org.

apparently worked hard in metaphysical circles to learn to understand and even to practice magic, so it would be almost unnatural to expect him to give away secrets. He is quoted as having told a friend that next to his poetry, "magic was the most important pursuit of his life." [241] As we will see shortly, Yeats makes his readers work a bit to get at veiled messages entrenched in his prose; for the unaware, it is possible to read his essays and miss the mystical connections.

About clairvoyant, mystic Helena Blavatsky who co-founded the Theosophical Society embraced by Yeats volumes have been written, but here it is interesting to relate a small story about her with regard to roses. She is said to have had an occasional awful temper; one such time was while she was in India in the presence of a particularly offensive chauvinist on whose head she made a shower of roses fall.[242]

The Hermetic Order of the Golden Dawn's link to the Rosicrucians is not at all clear, but there is apparently a connection if no more than in overlap of membership. The order has always embraced the rose in all its power and beauty and used it one of its symbols. Today the Rosicrucians are less hidden and there seem to be several different groups. One of the groups has built an impressive network supporting numerous publications, home study courses, a park in San Jose California, a Rosicrucian Egyptian Museum and even a Rose-Croix University International offering in-depth courses that expand on aspects of Rosicrucian studies, including

241 R. Cavendish, *A History of Magic*, p.146.
242 Colin Wilson, *The Occult*, p. 416-417.

the universal laws of nature. The University's purpose is to offer students opportunities for personal development and spiritual growth in a classroom environment under the personal instruction of faculty members who are considered experts in their fields of instruction. Yeats had more than a cursory understanding of the rose as a symbol; he knew its full meaning and connection to cosmic consciousness. Seward's analysis astutely took note of the difference in the tone of Yeat's writings after his initiation.

About the Rosicrucians not a lot was generally known during Yeats time and because several groups exist there is some confusion about which is which. Initially, the organization a great need for secrecy so as to keep its work from becoming adulterated, but enough seeped over the transom that we can be certain that a rose cross symbol was essential in its ceremonies. The initial Rosicrucian manuscripts were said to be first circulated in Germany around 1610; generally the society was begun "to afford mutual aid and encouragement in working out the great problems of life and in searching out the secrets of Nature; to facilitate the study of the system of Philosophy founded upon the Kabalah and the doctrines of Hermes Trismegistus."[243] An authority on the Rosicrucian Order, Paul Foster Case has explained that the name lives on and even flourishes because people are attracted by what they've heard about the Order and what they think they know of the early manifestoes which are actually very short,

---

243 Thomas D. Worrel, "A Brief Study of the Rose Cross Symbol," publisher and date unknown, p. 3.

but not intended to "move gross wits."[244] The Order is often misunderstood but does conceal itself from those who are incompetent. Paul Foster Case reminds us that the Order is not an organized society like the Freemasons and membership is not based on paying entrance fees, making an application or participation in a ceremony. "The Rosicrucian Order is like the old definition of the city of Boston: it is a state of mind. One becomes a Rosicrucian: one does not join the Rosicrucians," according to Case.[245] A contemporary booklet published by the organization today underscores that this is still the case. The group is said to "encourage open minded questioning and self-mastery" and to promote a higher "way of life."[246]

However we make sense of the foundations of the Rosicrucians or of the Hermetic Order of the Golden Dawn, a rose cross is used as a symbol by both fraternities and surely must have provided the essential connection to Yeats's psyche and the metaphysical messages he intended in *The Secret Rose*. First and foremost, the rose has been used as a "sign of silence and secrecy." In his "Brief Study of the Rose Cross Symbol," Thomas D. Worrel explained "the rose, like the cross, has paradoxical meanings." It is at once a symbol of purity and a symbol of passion, heavenly perfection and earthly passion,

244 Paul Foster Case, *The True and Invisible Rosicrucian Order: An Interpretation of the Rosicrucian Allegory and An Explanation of the Ten Rosicrucian Grades.* Weiser Books, Boston, 1985, p. 4.
245 Ibid., p. 5.
246 Mastery of Life, publication of the Supreme Grand Lodge of the Ancient & Mystical Order Rosae Crucis, AMORC, Inc. p. 7.

virginity and fertility, death and life." [247] What's more the rose has significance in numerology. According to Worrel, it represents "the number 5 because it has five petals. And the petals on roses are in multiples of five. Geometrically, the rose corresponds with the pentagram and pentagon."[248] In the Rosicrucian teachings, the "number 5 represents Spirit and the four elements."

But why did the Rosicrucians and other related groups like the Golden Dawn even bother to exist? What was the reason they came on the scene in the first place? Lynn Picknett and Clive Prince in The Templar Revelation provide an excellent interpretation of the whys and wherefores of the Order. The Rosicrucian movement and teachings were a major threat to the Catholic Church. The "idea of mankind's essentially divine status did not accord with the Christian idea of the 'original sin'—the idea that all men and women are born sinful because of the fall of Adam and Eve." [249] Thus the Rosicrusicians went underground, became secret out of necessity not because they wanted to operate in secret per se. Thus the Order was fully established long before the movement was noticed in the seventeenth century; "in fact it is scarcely an exaggeration to say that Rosicrucianism was the Renaissance."[250]

---

247 Worrel, "A Brief Study of the Rose Cross Symbol," p. 2.
248 Ibid., p. 2
249 Lynn Picknett and Clive Prince, *The Templar Revelation: Secret Guardians of the True Identity of Christ,* Touchstone, New York, 1997, p. 134.
250 Ibid., p. 135.

The Golden Dawn embraces both the lotus flower and the rose in its symbols. The Rose Cross Lamen is used prominently in consecration ceremonies and is worn at all meetings of Adepts. This symbol is so sacred to the Order that the instruction is given that it should never to be touched by persons.[251] The Rose Cross of the Order of the Golden Dawn is intricate and magnificently colorful; each part of it has profound meaning, but roses---rings of them—are at the center of the cross and each circle of them has meaning related to the elements, to the ancient seven planets and to the twelve signs of the zodiac and correspondence to the Hebrew alphabet. So the influence on Yeats's *Secret Rose* is clear.

What's more, studying Yeat's short story "The Secret Rose" carefully, it is almost impossible not to see parallels to the crucifixion of Jesus Christ, or to even more recently to the tragic death of well-known spiritual leader Dr. Martin Luther King, Jr. Both died ultimately, prematurely, in one way or the other—as does the protagonist in *Secret Rose*, for "stirring up the people" toward truth to use a phrase from Yeats. Seekers of truth are generally no strangers to dissociation and dissonance in their connection to non-seekers although it is their aim to connect. The disjunction with others more than likely results from the elevated frequencies that occur when one is seriously on the path of truth and higher awareness.

---

251 Israel Regardie, *The Golden Dawn: A Complete Course in Ceremonial Magic;* Four Volumes in One, (St. Paul, Minnesota, Llewellyn Publications, 1986) p. 310.

Yeats published "The Secret Rose" in 1897. Recalling the piece simply, outside of Seward's analysis and in more detail, it is an odd, short piece about an outcast named Cumhal who as a bard was doing what he did best: travel from hamlet to hamlet in a "particoloured doublet" singing and reciting poetry. Known in Ireland of the time as a gleeman, young Cumhal described his own soul as "indeed like the wind, and it blows me to and fro, up and down and puts many things into my mind and out of my mind, and therefore I am called the swift, wild horse."[252] Yeats paints the portrait of Cumhal as an outcast, one who managed late one evening to so offend a certain friar named Coarb that he decided to kill him. The friar's reasoning was clear as to why the poet should be put to death by crucifixion: "the bards and the gleemen are an evil race, ever cursing and ever stirring up the people, and immoral and immoderate in all things, and heathen in their hearts."[253] Obviously, Yeats paints the friar with an inability to see himself: full of mean spirit, hateful and swift to judge as he prepares to complete a bleak, horrific act.

The atrocious friar determined quickly that crucifixion was the appropriate punishment for the gleeman; and so Coarb even managed to awaken fellow monks from their sleep to seek assistance in overseeing the construction of a cross on which to hang the gleeman. Friar Coarb wanted to spread meanness around a bit; it wasn't enough to contain it

---

252 James Pethica, ed. *Yeats Poetry, Drama and Prose* "From *The Secret Rose,* 1897" Norton Publishing, New York, 2000, p. 195.
253 Ibid.

within himself. But all during the construction of the cross, Cumhal seems undaunted, not fearful, still a free spirit, for he knows something the friar and his accomplices do not know. The gleeman's being is entrusted to the Holy Spirit, as Yeats reveals so subtly in the metaphor of a rose: "...I have been the more alone upon the roads and by the sea, because I heard in my heart the rustling of the rose-bordered dress of her who is ...subtle-hearted and more lovely than a bursting dawn to them that are lost in the darkness."[254] In the foregoing lines, the rose-bordered dress is indeed, metaphorically, the Holy Spirit. Unbeknownst to the friar and his accomplices, Cumhal has already made the intimate connection. Prepared and unafraid to die an earthly death because he knows something profound, Cumhal already has everlasting life.

In his selection of character and plot, Yeats assuredly confronts and affronts ancient, pervasive problems of the Christian Church and indeed a number of world religions: deceit, violence, hypocrisy and egoism. The friar, showing no mercy, was very quick to judge the gleeman; so quick in fact that he could not wait until morning for fellow clergy to awaken so he could influence them to put Cumhal to death. In one swift, bold move, as Yeats paints him, the friar makes a 180-degree turn about from the real, true teachings of Jesus Christ of the Christian Bible. The friar swiftly steps over of the tenets of the Christian faith: forgiveness, love, charity and tolerance to determine that the young bard should die

254 Ibid. p. 197.

immediately simply because of ill words toward the Church. Yeats sets up his characters provocatively to make a profound criticism of clergy and, since the monks acted in collusion, the Church itself.

In "The Secret Rose," Yeats' selection of adjectives such as "swift, wild horse" is pregnant with metaphysical meaning. The gleeman was not afraid to be crucified because he was safely in a connection with the one and only incorporeal God or what those in the metaphysical tradition would refer to as endlessness or cosmic consciousness. Moreover, Yeats describes the gleeman as having a "bulging wallet" another metaphoric indicator that he was full of what Christians commonly refer to as "Holy Spirit." And thus the story goes on to a scene leading up to the crucifixion when Cumhal gives away what is the equivalent of his last supper; the gleeman throws strips of bacon among beggars who fought until the last scrap was eaten. Yeats gives his main character, Cumhal very few last words to moan in the midst of the grumbling beggars before he dies. "Outcasts," he calls to their unheeding ears, "have you turned against outcasts?"

More than six centuries before Yeats, and nearly a century before Dante used the rose in carnal and spiritual symbolic manifestations, Mevlana Jelaluddin Rumi an evolved Afghanistan- born spiritual writer made the same provocative flower an essence in his poetry. Born on September 30th 1207 in Balkh when Afghanistan was still part of the Persian empire, Rumi was instructed by his father's secret inner life and became a sheikh—a religious scholar who helped the poor.[255] But the pivotal event in his life seems to have been meeting the enigmatic saint Shamsi Tabrizi, his teacher and spiritual guide.

As a result of the remarkable translation work of Coleman Barks, Kabir Helminski and others we can now experience Rumi's delightful, spirited poetry firsthand. Barks and his team translate so smoothly that it is almost as if Rumi wrote originally in English. Or is it the case the Rumi's poetry is so spiritually profound, so inspired that the intended messages cannot be missed in translation? For me, the resonance of Rumi's poetry seems to transcend all earthly language. The poems reverberate even in translation although they were not originally written in English. The lines are so pure that they seem palpably connected to Spirit, it seems as if Rumi's poetry was *intended* to be translated and circulated in a variety of global languages. Coleman Barks' collection, *The Essential Rumi* is simply a splendid capturing of Rumi's allegorical and spiritual poetry as it was meant to be. Witness for instance Rumi's "Spring Is Christ." The lines "We walk out to the garden to let the apple meet the peach, to carry messages between rose and jasmine"[256] clearly signify connections between the carnal and spiritual as do the next several lines that use the rose as a metaphor for reaching "a lamp"—of course the lamp is the "Holy Spirit":

255 Coleman Barks et. el. Translators, *The Essential Rumi*, Castle Books, Edison, New Jersey, 1997, introduction.
256 Ibid. p. 37.

Spring Is Christ
Raising martyred plants from their shrouds
Their mouths open in gratitude, wanting to be kissed.
The glow of the rose and the tulip means a lamp
Is inside...[257]

There is more than one use of the rose in Barks's collection, but it is Rumi's "Dying Laughing" that seems most poignant in its use of the rose in metaphysical symbolism. The poem opens with a lover who told his beloved how much he loved her and how "faithful and self-sacrificing" he was toward her. "There was fire in him. He didn't know where it came from..."[258] But his beloved was apparently not impressed with all the material things he'd done and she told him as much. She remarked to him in essence that he had done very well as a lover in outwardly acts but he hadn't died. To hear that he had not died struck him as funny since he'd done so very much for his intended.

When he heard that, he lay back on the ground
laughing, and died. He opened like a rose
That drops to the ground and died laughing.[259]

It is when the lover died laughing, a metaphor for letting go of the material world that he "opened like a rose" that is

257 Ibid. p. 38.
258 Ibid. p. 212.
259 Ibid.

to say he made the quantum leap toward the Holy Spirit. The point is that again and again the poet Rumi shows us that he knew and understood everything necessary to make the ultimate connection, long before Dante's epic was written.

Not too far a field from the character Cumhal in Yeat's "Secret Rose" or the lover in Rumi's poem who learned to "die laughing," the old man in Hakim Sanai's poem "The Old Man of Basra" lived in a constant struggle to make the connection with the Holy Spirit. The old man arises daily to the same mundane, unessential questions that millions of human beings worldwide ask daily "What shall I eat?" and "What shall I wear?"[260] The old timer lived in constant resistance to such questions when they plagued his psyche. To the question what shall I eat, he learned to answer "death," and to the question, what shall I wear, he learned to answer "a winding sheet." These answers revealed spiritual evolvement; the old man of Basra was fighting to overcome the material world. Sanai, writing nearly a hundred years before Rumi, used "garden roses" oddly, in juxtaposition, to show that for "self-Cherishers" roses assume the form of "malignant boils." This is the case for those who are imprisoned by the three prisons of "deceit, hatred and envy" reminiscent of the three beasts encountered by Dante in the *Divine Comedy*. Sanai, about whom little is known, lived during the reign of Bahramshah (1118-1152) and probably died in 1150.[261] His

---

260 Hakim Sanai, *The Walled Garden of Truth*, Translated and Abridged by D. L. Pendlebury, (The Octagon Press, London, England) p. 7.
261 Ibid., p. 37.

work is little known in the West, but has a deeply metaphysical quality to it.

Thus not only Rumi but Sanai knew eons before Dante and Yeats precisely what the latter came to know as a result of deep study of the ancient teachings. Connection to the Light of the Source takes a lot of study and work done in gradations. Dante Alighieri was guided by allegorical messages in St. Jerome's Latin Vulgate version of the *Holy Bible* that he likely studied and mastered. In his epic poem, he astutely uses Beatrice as a guide; she had rebuffed him in real life. Thus, in an interesting juxtaposition, Alighieri shows that he has overcome the material world. Yeats, in his association with the Theosophical Society and the Order of the Golden Dawn, we can be sure is likely to have had teachers or guides to hermetic knowledge; Helena Blavatsky was surely one of them. Yeats in his connection to several esoteric organizations had many opportunities to become privy to hidden knowledge that has been kept out of the reach the profane masses. He was fully involved in metaphysical studies and had reached a high level of understanding about realms that are far beyond the material world. It has been said that when the student is ready, a teacher will appear. For Rumi, Tabrizi arrived when he was ready; for Sanai we can't be sure since he lived so long ago and not much has been written about him, but it appears that his teachers were Lai-Khur and Yusuf Hamadani.[262]

---

262 Ibid., p. 8.

One has only to look at the meanness and lack of humility in the world today to understand why mystics and adepts were careful not to waste time on "gross wits" and on those who refused to do the necessary work toward what mystically based secret societies refer to as "self-mastery."[263] What all of these great writers and poets from Yeats to Rumi apparently knew well is that human beings are supposed to mimic the rose in their behavior in order to trump material world.

---

263 Hidden knowledge and the secret teachings of Jesus are mentioned and alluded to throughout the *Bible*, including in Mark 4:11 which says "Unto you it is given to know the Kingdom of God: but unto them that are without, all these things are done in parables." That is to say many people worldwide read the Bible, but very few fully understand it. It is a tremendously complex book that takes years of study, reflection, and pure intent to understand. This too is made clear in the Book of Mark, Chapter 4, Verse 12: "Seeing they may see and not perceive; and hearing they may hear and not understand; lest at any time they should be converted and their sins should be forgiven them."

9

Rose Vignettes:
Black Plague to Gulag

By Frankie Hutton*

In 1938, German-Jewish poet and writer Rudolf Borchardt published one of his last works, a masterpiece of a book under the title *The Passionate Gardner*. In this little known work of about 340 pages that has been translated to English by Henry Martin, Borchardt uses plants and gardening as a deep antecedent for the potentiality of human beings to grow up, to master themselves and to create something good for all of mankind. Borchardt beseeches us to understand a "grandiose metaphor in which the human soul is itself a garden, strictly and fervently in the care of God: the heart of the child of the world is rife with weeds, and He plucks them up; the heart of the innocent is a pious modest field full of silent grace, and fragrant with fine perfume..."[264]

It is no coincidence that the *Holy Bible* opens with the scene of a garden. Borchardt reminds his readers cogently

---

[264] Rudolf Borchardt, *The Passionate Gardener*, English Translation by Henry Martin (Kingston, New York, McPherson Publishing, 2006) p. 8.

that the scene in which Christ took his pained departure from the world also took place in the garden of Gethsemane. Thus, "purity can only be found within the protection of vegetation and everything which is not impure is a garden." The garden then ultimately is a place of order, where man is potentially master and transformer. Borchardt used the garden and flowers as a metaphor for the great work that human beings can accomplish when they are able to transcend the ordinary, to open up, to bloom like so many flowers. By this view, there is significant merit in pure work that leads to creation of something good for all of human kind. Constance Casey, who reviewed *The Passionate Gardner* for the *New York Times* in 2006, refers to Borchardt's work as a form of "botanical globalism" that points us to the heart of an urgent matter in gardening: "whether to grow native plants only, or to permit plants from other continents in our gardens." Ultimately gardeners can make the choice of whether to reach out to embrace plants that are uncommon to the area or to settle for provincialism—that is to stay with plants and choices that are common to their geographic areas. This metaphor from gardening is easily applied to service to humanity. Are we to live separately, selfishly and just for our own families and immediate circles or do we swing outside that realm to create something wonderful and transcendent for the good of all? In this regard, it is important to underscore the fact that the rose lends itself to cross breeding more than any other flower and also renders a number of gifts to humans, including beauty, medicinal qualities, fragrance and perfection in its very essence. In this sense, the rose services humanity as something beautiful and useful. The choice of staying comfortably within our own spheres or swinging out of the familiar to connect with, to help, and to support others and to make a difference through creating something wonderful that all can benefit from is available to all of us. A number of great human beings have understood Borchardt's powerful metaphor of botanical globalism and have reached far beyond themselves, often through great adversity, to create something good, beautiful and useful that advances or aids human beings. This is the acid test, to move away from our own trials and tribulations to do something for others: to work intently, to serve others for no gain to ourselves, no matter what the adversities. Very difficult work indeed; some of us cannot even understand why it is paramount.

Borchardt's powerful insights and metaphors of gardens and flowers point us poignantly to individuals who have embraced the rose in a number of manifestations, including healing, music, social justice and world peace for the rose is quite singularly a symbolic presence in all of these. When it comes to metaspiritual matters, the rose is a remarkably useful symbol of the opening of the chakras, those urgent, unseen energy centers in the human body that average human beings never come to know. When we consider the vastness and universality of the rose, the flower no doubt reaches its earthly paragon or highest level when Daniel Andreev put it forth as a potent symbol of all people uniting in one deeply moral force for world peace. We will meet Andreev in this chapter and others who like him who have known the superior qualities of the rose in their work, often as a result of a strong spiritual connection or light that revealed itself quite vibrantly when one is ready.

The extraordinary beings who have understood the utter necessity for work that aides others and so eludes most of us have traversed this earth in diverse geographic locations in practically all epochs and have made their marks in most legendary, extraordinary ways as will be explored in this essay. Given the focus of this anthology, who would be surprised to know that the work of these individuals in some way embraced the rose, not only symbolically, but also practically since the botanical essence of various roses has been found to possess magnificent medicinal properties.

Dr. Edward Bach

So let us consider healing and medicinal properties of the rose first as understood and used by Dr. Edward Bach (1886-1936). Bach, a British physician who studied medicine at the University College Hospital, London, found plants as a curative after abandoning his traditional, successful practice. Bach's work with plants was antithetical to the traditional Western approach to medicine that was the foundation of his medical training. He is said to have come to the understanding of plant essences psychically. The flowers used by Bach are thought to be of a "higher order" since they contain frequencies that are within the human energy field.[265] This is perhaps a novel idea for most of us who have never considered flowers to be more than lovely to look at, sometimes to smell or to adorn tables and parlors. The human soul is thought to contain 38 qualities or virtues that are the same qualities of flowers selected by or "given to" Dr. Bach from a higher realm. The energy or divine sparks of these flowers are like wavelengths in the energy field that Bach came to know intuitively.[266] Dr. Bach introduced the essence of wildflower blooms, including rock rose and wild rose to treat all sorts of ailments, including asthma, irritable bowel syndrome, headaches, muscular tension, rashes and more. Bach flower essences, a spin off of the ideas of homoeopathy but said to be more pure because they lack the products of the disease, are now carried in top health food stores worldwide and are also available through mail order. The essences can also be used to treat mental upsets such as remorse or lack of confidence. Based on these 38 powerful flower essences, Dr. Bach's work is also akin to herbal medicine, but has as its foundation his thought that "disease is in essence the result of conflict between Soul and Mind."[267] Bach reminds us that although bacteria plays a part in the spread of physical disease, it is only part of the equation because not everyone exposed to bacteria gets sick or develops disease; there are factors

265 Mechthild Scheffer, *Bach Flower Therapy* (Rochester, VT, Healing Arts Press, 1988), p. 16-17.
266 Ibid., p. 16.
267 Edward Bach and F. J. Wheeler, *Bach Flower Remedies* (New Canaan, CT, Keats Publishing, Inc. 1997), p. 7.

higher than the material world that must be considered.[268] In his essay "Heal Thyself," Bach explained that disease serves the purpose of letting us know our essential faults and if we can correct the fault, the disease has been beneficial and will be alleviated:

> No effort directed to the body alone can do more than superficially repair damage, and in this there is no cure, since the cause is still operative and may at any moment again demonstrate its presence in another form. In fact, in many cases apparent recovery is harmful, since it hides from the patient the true cause of his trouble, and in the satisfaction of apparently renewed health the real factor, being unnoticed, may gain strength.[269]

Throughout recorded history, a few extraordinary people like Edward Bach have mastered themselves and the feat of connecting with others for the purpose of advancing work that aids humanity. Ironically, as we are beginning to see, all of these individuals from countries and cultures wide and far have in some way used or embraced the rose in their work.

268 Ibid.
269 Ibid., p. 10.

Nostradamus, 16th Century France

Although there were detractors and disbelievers all around him, the great Michael de Nostradamus led an extraordinary life and his powers seemed to increase as the years progressed. From the universe, Nostradamus was given the unusual, bitter-sweet gift of prophecy. His work and predictions have not gone unnoticed in the past five hundred years for a number of films, books and articles have chronicled his life and work. More than an astute physician, clairvoyant and mage, the Frenchman known simply as Nostradamus predicted the explosion at the World Trade Center in New York and numerous other devastating events over the past centuries. Destined to be as controversial as he was sought after for his psychic powers by the rich and the royal, he was conversely thought by his wife's family to be a failure for his inability to save his own family from the Black plague when it struck Agen in southern France of 1537. For this, he was asked to return the dowry paid by his wife's kin at the time of their marriage. An Italian doctor and apparently jealous friend, Julius-Cesar Scaliger (1484-1580) provoked a bitter quarrel between Nostradamus and his in-laws which shattered their friendship; this hurtful time was followed by a charge of heresy and exile from Agen.[270]

---

270 John Hogue, *Nostradamus: New Revelations* (Rockport, Massachusetts, Element, 1994) p. 16-17.

Ian Wilson's fine biography of Nostradamus provides an excellent translation from the treasurer of Aix-en-Provence when the plague hit that town in 1546, less than a decade after it struck Nostradamus's family. Victims suffered greatly from "Black egg-sized swellings or buboes would appear in the armpits and groin, oozing blood and pus. Black blotches would follow all over the skin accompanied by severe pain and within five days the victim would usually be dead."[271]

Although there was said to be no known cause and no known cure, the conditions ushering in the plague were a combination of strange weather patterns and unsanitary living habits that left Europe ripe for an epidemic of some sort. Peter Lemesurier, a Cambridge-trained linguist and teacher has helped us to put the foundation causes of the plague in perspective. By the 1520's, Lemesurier explains how the conditions for the plague started while "Europe's Little Ice Age was in full swing. As a result winters were tending to become more artic, summers colder and wetter. Even when plowing and sowing could be done at all, the crops would often rot in the fields. No crops meant no food, and no food meant no survival."[272] As a result of this strange weather pattern, France suffered agriculturally and experienced "no fewer than thirteen major famines during the course of the century."

But Europe also suffered flooding as a result of strange weather patterns. It is clear that odd weather patterns aided

271 Ian Wilson, *Nostradamus: The Man behind the Prophecies*, p. 50.
272 Peter Lemesurier, *The Unknown Nostradamus* (John Hunt Publishing Ltd., 2003) p. 62.

and abetted flooding and that coupled with trauma to the agricultural system was a mix for disaster. By 1544, heavy rains caused the Rhone River to rise beyond normal levels causing flooding of towns and cemeteries; corpses washed out of their graves and exceptionally awful sanitation problems led to the plague.[273]

Although the predictions of Nostradamus and even his life's story have been well recorded, information about his clever use of rose petals in a mixture as a preventative during the years of the plague is scant. Ian Wilson interprets and recounts Nostradamus's own *Treatise on cosmetics and jams* that prescribes use of the rose as a preventive for those who had not yet caught the plague.

> Take some sawdust or shavings of cypress-wood, as green as you can find, one ounce; iris of Florence, six ounces; cloves, three ounces; sweet calamus [cane palm], three drams; aloes-wood six drams. Grind everything to powder and take care to keep it all airtight. Next take some furled red roses, three or four hundred, clean, fresh and culled before dewfall. Crush them to power in a marble mortar, wooden pestle.[274]

Nostradamus then advised adding more half-unfurled roses to the mixture and rose juice as well. He advised further

---

273 Ian Wilson, *Nostradamus: The Man behind the Prophecies*, p. 45.
274 Ibid., p. 47.

that the mixture should be made into pill form and that it would smell good too. What the great mage did not explain and may not have known in the scientific-medical language of his time is that rose petals contain important phytochemicals such as beta-carotene, botulin, catechin, flavonoids and nutrients such as calcium, iron, magnesium, manganese, phosphorus and potassium; the petals also contain an array of B vitamins and is generally good for infections of all kinds.

Dr. Tomin Harada, World War II Era, Japan

Few in America or Europe outside of medical profession have heard of Tomin Harada, for before World War II he was just another professional serving his country as he was trained and expected to do as a military physician. The World War II era changed all of that. Born in 1912, Harada graduated from Jikei Medical School in Tokyo in 1936 and soon after in 1938 was trust into war with China, one of the opening stages of World War II. Dr. Harada put himself in a tense situation as a military officer-surgeon when he was asked to talk to a group of soldiers using a canned model prepared for him for such occasions. "...I just could not bring myself to pass on such nonsense."[275] So Harada decided to speak from his heart and remembered his speech to be something like this:

275 Tomin Harada, *Hiroshima Surgeon*, (Newton, Kansas, Faith and Life Press, 1983), p. 4.

Although the Japanese and Chinese races are slightly different, we along With the Tungas people, the Koreans, and the Melanesians, all have Chinese Blood flowing in our veins. More importantly, Japan has received much from Chinese culture. We are now fighting, but we must not let this become a war Of hatred. The time will surely come when we will be friends again. We are going now to help create an opportunity for that friendship to be reawakened.[276]

Given his officer training, Harada was embarrassed by the speech later after he thought about what he'd said but he had not been able to bring himself to say things he didn't believe. He had been critical of Japan's notion that it should rule the world just has he resented the colonialism of Great Britain, the United States and Holland. Stationed at the Taiwan Henchun garrison amid poisonous snakes and mountain leeches, Harada, then practically emaciated from the long campaign with few rations left, remembered the day the news came that the war was over. He also found out shockingly that Hiroshima and Nagasaki had been attached by Atomic bombs. Eventually returning to his hometown of Hiroshima, he found it obliterated and his family missing. With the aid of a friend who survived, Harada eventually

276 Ibid.

found his family at their ancestral home outside of town and soon after set himself to work building a hospital in 1946 to help the survivors of the bomb's devastating impact.

It was after the war and the death of his wife that Harada became a passionate peace activist and in retirement promoted growing roses in connection with peace activism. He met another former WWII officer in the UK who introduced him to roses. Harada immediately took to the lovely, fragrant flower and became a breeder; both men agreed that roses could be used as a symbol of world peace. It is noteworthy that two former officers who had seen the horrors of war from opposing sides could now embrace the simple, fragrant rose as a symbol of world peace.

Dr. Harada's superior reconstructive, plastic surgery in the treatment of the bombing victims in Japan and later his treatment also of those who suffered maladies of the herbicide agent orange and napalm used in the Viet Nam War were progressive and legendary. He is said to have been especially attracted to roses because the fresh flower reminded him of the young and beautiful children who had been killed in the war and would never have the opportunity to lead the world toward peace. In crossbreeding roses, Dr. Harada developed new brands that he gave unusual names such as "Phoenix Hiroshima, Hiroshima Mind, Miss Hiroshima, Peace Maker, Red Hiroshima, Dr. Tomin, May Peace, Spirit Hiroshima…" and so on that still exist in Hiroshima and in select gardens all over Japan.[277] During his life time, Dr. Harada attempted to send some of the flowers he bred to world leaders with the wish for no more nuclear war and a "free" world. Dr. Harada explained, "Let the roses speak for themselves. Roses embody peace and beauty by themselves." His rose initiative caught on in the corridors of Japan and perhaps the peace movement in that country today is a spin off of his untiring work with roses for peace. He died in June 1999.

Mozart (1756-1791), Austria

The rose is a powerful symbolic presence in the universe, but its connection to semiotics is easy to miss for the profane and noisy probably because the flower itself is at once quiet, beautiful and refined. Likewise high vibration beautiful music seems rarely to be enjoyed by the profane. Song writers and composers all over the globe have embraced the rose in exquisite compositions and it could be deduced from a couple of striking examples that at least some of those who have used the flower were no strangers to higher awareness. This is certainly the case with Mozart. When his life and work are carefully studied, it is apparent that he knew a lot about higher awareness and mysticism and there may have been resentment from colleagues that he revealed

---

277 Facts provided by Hisae Ogawa in translation from Dr. Harada's Japanese writings to English from: *Moment of Peace* (date unknown), Gariver Products Co, Ltd. *Hiroshima Roses* (1989) Mirai-sha Publisher; and *Chasing the Dream of Peace* (1983), Kei-shobo Publisher, Japan.

it. Thus speculation continues that breaking silence about secrets revealed to him in Masonic initiation may have led to his mysterious, untimely death.   Wolfgang Amadeus Mozart as a child prodigy mastered the violin, harpsichord and organ and went on to compose some of the greatest classical music of all times. The evidence is that Mozart during this short life knew the mystic rose. One of his greatest renditions, *Die Zauberflote* or *Magic Flute* mostly finished between 1790-91 just before he died, speaks volumes about arcane messages of the rose. The opera was written to convey important Masonic and mystical messages, although that fact is apparently grasped by only a few who have seen and heard the opera worldwide. Naturally, the rose adumbrates in this most widely known work, one that is rift with secret symbols. Important tenets of Freemasonry are service, silence, patience, steadfastness: these are replete in the *Magic Flute*. It is not by chance that Judy Taymor's breathtaking production of the *Magic Flute* at New York City's Metropolitan Opera during the 2007-2008 seasons prominently depicts a rose garden in act two, scene three. Taymor, an award winning, renowned theatrical producer, was so astute as to make the roses lighted, crystal, red and colossal in size. The rose symbolizes love, discretion, and the potential for connection with the Source of all and everything; it is truly the mystic rose. Both the garden and the rose are potent symbols that can lead to the light of Cosmic Consciousness—that is to say, to becoming one with the Source, for those who are aware and not blinded by deceit, ego, and greed. This truth has been veiled for hundreds of years while various denominational divides were purposely created to keep the masses better managed, manipulated and placated.

In *The Magic Flute*, protagonists Pamina and Tamino, lovers in the opera, overcome evil in their midst and grow tremendously in their understanding that true love is unconditional and never too possessive. These are most difficult lessons for Pamina who does not comprehend Tramino's silence during his initiation to higher consciousness. She lacks Tamino's discipline and resolve to stay the course, to become a master, first and foremost of himself. With the aid of three spirits, Pamina eventually finds Tamino and walks with him through many ordeals symbolized in the opera by trials connected with water and fire. Through it all they are protected by a magic flute. They are united in marriage at the end of the opera, but this too is a very, most powerful and misunderstood symbol in mysticism.

Now this marriage that takes place between Pamina and Tamino in *The Magic Flute* is important to consider closely for it teaches us much about the real meaning of marriage as it is practiced worldwide. "Marriage and intercourse, whether legal or illicit, refer not to any carnal relationship but to a spiritual "marriage", or blending of consciousness, at any level.[278] Thus, as we come to understand that the practice of marriage on earth is really symbolic of the potential of the human being to rise up, to fuse with the light that transcends

---

278 Geoffrey Hodson, *The Hidden Wisdom in the Holy Bible*, Volume I, (Wheaton, Ill. The Theosophical Publishing House, 1963) p. 129.

darkness. The ultimate goal always is to connect with and become one with Cosmic Consciousness. When individuals marry, a fusion or uniting is supposed to take place such that the two rise up to become more than each one could have been solo, but often that does not occur in our troubled world. This knowledge that marriage is intended to *symbolize* all that is sacred is urgent and important for us involved in the ultimate earthly relationship of matrimony to keep in mind. Most of us have not been socialized to understand what the institution of marriage truly signifies, but the troubled practice continues worldwide. Even when the truth of the institution of marriage has been made available in metaphor and symbols such as the rose and the cross, human beings still do not get it: intense work on the part of both partners is involved to make successful marriages on earth and likewise intense work is necessary to reach the light of the Holy Spirit. Rather than doing the intense work, the preference of human beings has been to intellectualize the real messages provided in scripture through talk. Egos abound too and get in the way of successful marriages on earth; it goes without saying that too much ego is a impediment to connection to Cosmic realms just as it impedes successful relationships in our daily lives. Commitment to the work associated with rising up has to be great, so much so that most human beings cannot or will not discipline themselves to perform it. Likewise the institution of marriage, like sex between a man and a woman, are simply not routine matters. Inherent in these earthly relationships of sex and marriage is the truth that we as human beings all are intended to connect with each other, to love and to be supportive to one another and ultimately to connect with the Source of us all. We are not on earth to war with each other. We may never understand this essential truth. Great efforts have been made to block it from our understanding.

Charles Austin Miles, New Jersey, USA

Little is known about American Charles Austin Miles (1868-1946) at this remove. Most people today have never even heard his name uttered but practically all in the protestant Christian tradition know his song "In the Garden" as a staple of Sunday church services. The song was "given" to Miles in 1912 as a result of a request for him to write a hymn text that would be "sympathetic in tone, breathing tenderness in every line; one that would bring hope to the hopeless, rest for the weary, and downy pillows to dying beds."[279] Miles was guided in writing the song by a vision of Jesus preceded by opening his Bible to John 20, the meeting of Jesus and Mary. Seated in a dark room at the time, Miles' account of writing the famous song has been described in *101 Hymn Stories* just as he remembered it:

I awakened in full light, gripping the Bible, with muscles tense and nerves vibrating. Under the

---

279 Kenneth W. Osbeck, *101 Hymn Stories*, (Grand Rapids, Michigan, 1982), p. 124.

inspiration of this vision I wrote as quickly as the words could be formed the poem exactly as has since appeared. That same evening I wrote the music.[280]

Like the writer William Butler Yeats whose work embracing the rose was discussed in the last Chapter, Charles Austin Miles had Masonic connections as mentioned in his 1946 obituary in the *New York Times*.[281] Miles was born in Lakehurst, New Jersey and was the author of 3,000 hymns. "In the Garden" was not considered a financial success in Miles's lifetime, although it sold three million copies and recordings exceeded one million. The hymn has worldwide distribution and is sung in many languages.[282] We can only speculate that he have been a member of one of the secret societies such as the Rosicrucians or even the Order of the Golden Dawn since both fraternities embrace the rose fully. Miles was precise in his lyrics embracing roses in a garden for the composition of *In the Garden*. The dew on the roses Miles refers to is symbolic of light. Obviously, Miles knows something about Cosmic Consciousness; he had apparently tapped it based on his description of how the song was "given" to him.

280 Ibid.
281 *New York Times*, March 12, 1946, p. 26.
282 Ibid.

I come to the garden alone
While the dew is still on the roses
And the voice I hear falling on my ear
The song of God discloses

In biblical hidden teachings, a garden is symbolic of paradise and by now it is essential that we also know that the red rose when fully opened symbolizes connection of the key crown charka to what has been synonymously called cosmic consciousness, the universal mind, Christ consciousness or God. It is also clear that each of us must make the journey to the Source alone, thus the opening line of the song, "I come to the garden alone."

Daniel Andreev

In his own realm—that is to say the mid-twentieth century metaphysical community in Soviet Russia, Daniel Andreev was the mastermind of an incredible work known simply as *The Rose of the World*. At the time that he wrote the masterpiece, no one in the West knew of it and very few in Russia knew of the work either because by 1947 Daniel Andreev had been sentenced to 25 years in a gulag for his outspoken ideas and other writings that were considered anti-Soviet. Andreev's pre-gulag work was destroyed; his opus *The Rose of the World* was written while he was in prison, hidden with the aid of those around him who thought the work worthwhile, buried until the collapse of the Soviet system, printed first in Russian in Moscow and now in English. And no wonder the harsh Soviet system repressed Andreev and his work. No words were minced about the evils of power hungry government officials by Andreev in *The Rose of the World*. He gets to the heart of heady political and economic matters that beg to be confronted by citizens all over the globe, now more than ever. In recent English translation by Jordan Roberts, Andreev's writings proffer the notion that human beings should be tremendously concerned by the formation of a global police state that he felt was in the making. Andreev's vision was that new technological advances simply helped along the process of a global police state. Andreev wrote sometime between 1947 and 1957:

> It should come as no surprise today that one side of every scientific and technical advance goes against the genuine interests of humanity. The internal combustion engine, radio, aviation, atomic energy—they all strike the bare flesh of the world's people with one end, while advances in communications and technology enable police states to establish surveillance over the private life and thoughts of each person thus laying an iron foundation for life-sucking dictatorial states.[283]

283 Daniel Andreev, *The Rose of the World*, English Translation by Jordan Roberts (Hudson, New York, Lindisfarne Books, 1997) p. 10.

It is eerie to recall Andreev's warnings now, almost fifty years later, considering American writer Norman Mailer's pronouncement just before he died in 2007 that the devil's principal weapon is technology. Mailer seemed to believe that the devil aspires to create a mechanized world where "souls are increasingly interchangeable."[284] Whether that's true or not, we cannot know at this remove, but it does seem clear that technology often hurts as much as it helps sort of like the Southern story of the cow who gives a splendid bucket of milk and then takes her hind leg and kicks it over. Human beings are too much steeped in usage of cell phones, computers, iPods and such to even notice or care that these are used as potent tracking devices that are an invasion of privacy. The breech of quiet for others caused by impolite cell phone use is out of control. Coupled with high tech weapons, all of this so called "technology" is clearly detrimental to all of us now and will be even more so in the long run when radiation is factored. Keen note must be taken of Andreev's admonishments about negative aspects of technology in light of recent televised Congressional hearings on the development of a new mercenary, security force known as Blackwater, USA. This organization is apparently an extraordinary group of super charged, highly trained men with the latest high tech weapons and technical know-how who will go anywhere and do practically anything at the bidding of government.[285]

---

284 Read Michael Lennon's interview with Mailer in the article "The Rise of Materialism," *New York Magazine*, October 15, 2007, p. 24-29.

285 Of interest is Jeremy Scahill's book *Blackwater: The Rise of the World's Most Powerful Mercenary Army* (New York City, Avalon Books, 2007).

Andreev was released early from prison after a decade of poor diet, beatings and harsh, cruel treatment that left him broken and very sick. Upon release, an emaciated Andreev knew that his days were numbered, yet he persevered to finish *Rose of the World* which he did just before his death. We are reminded by Mikhail Epstein who has translated and interpreted a large portion of Andreev's work that he was preoccupied with nature and this component drives much of the philosophical underpinning of his work. Epstein concluded:

> Throughout his creative years, Daniil Andreev suffered under the Pressure of official Soviet ideology's "stubborn iron materialism," but
> His inner resistance to this mysticism of materiia did not push him to the other extreme of bodiless spiritualism. Nature was the center of his Whole system, and he singled out a special category of "elementals"(*stikhiali*), spiritual entities that have an elevating effect on the human Soul and are embodied in such natural elements (*stikhii*) as rivers, trees, wind, and snow. Daniil Andreev enjoyed traveling through the wildest and most remote Russian forests, because for him nature suggested the most genuine way of knowing God...[286]

286 Bernice Glatzer Rosenthal, *The Occult in Russian and Soviet Culture* (Ithaca, NY, Cornell University Press, 1997) p. 336.

His marriage to Alla survived the gulag, but he didn't survive very long after release.   Alla Andreeva had suffered too because she was incarcerated in the same system, although at a different locations and away from her husband. Alla Andreeva remembered her time in the gulag as demeaning and harsh. The gulag system was apparently engineered to strip every ounce of individuality and dignity. Ann Applebaum's award winning book *Gulag*, refers to Alla as Anna Andreev and provides much insight into the horrors of the Soviet prison system. At first Mrs. Andreev was sent to a camp where prisoners were allowed to wear their own clothes, but beginning in 1948 she was made to wear a smock dress that she found particularly offensive.[287]

In order to counter the potential for a global government and attendant global police force that he felt were coming, Andreev advocated the necessity for a strong moral body. In *Rose of the World* seems almost to Andreev plead that:

> We must...recognize the absolute necessity of the one and only path: The establishment, over a global federation of states, of an unsullied, incorruptible, highly respected body, a moral body standing outside of and above the state. For the state is, by its very nature, amoral.[288]

287 Ann Applebaum, *Gulag* (New York City, Penguin Books, 2006) p. 177.
288 D. Andreev, *The Rose of the World*, p. 10.

The Global Anti- FGM Movement

Surely Alice Walker, Pulitzer Prize winning America author, did not anticipate being so roundly criticized and in some corridors practically skewered for her work that brought the horrific practice of female genital mutilation to the spotlight. Since disclosure is essential to the developing global movement to fight female genital mutilation, Walker's book *Possessing the Secret of Joy* was a bit of a blockbuster. After social activist, humanitarian, nurse Efua Dorkenoo read Walker's book, she immediately contacted Walker as she explained to colleague in the struggle Tobe Levin: "I wrote to Alice. You see FORWARD [the organization founded by Dorkenoo to fight the practice] counsels women like Tashi [the main character Walker's book] whose mental anguish at having suffered mutilation has become unbearable. Tashi is so real that I wanted to let Alice know and invite her to be patron of FORWARD. Of course, she agreed..." a FORWARD board meeting in June 1995 specifically credited Alice Walker's work in the long struggle to eradicate the practice. The film *Warrior Marks* was largely inspired and guided by African activists of Efua Dorkenoo's stature but Walker naturally received the lion's share of flack. After the film was shown in London, Dorkenoo even received death threats since she was a promoter of the film, an avid fighter against FGM and now lives in the UK. Walker continues to be criticized. In 1993, *Newsweek* quoted Dr. Nahid Toubia who said Alice Walker was a "writer whose star is fading...

trying to sensationalize in order to get the limelight back." Equally alarming, in a 1993 *New York Times* editorial, African residents in the United States blame Walker for their distress, reducing to personal display her attention to "their" issue. Despite an abundance of criticism, Walker, like Dorkenoo, has courageously continued her work and both have been propelled, not quieted.

Efua Dorkenoo, OBE, is a brilliant nurse and native of Ghana who moved to the UK in the 1970s and is one of the earliest campaigners against FGM. She founded FORWARD (Foundation for Women's Health Research and Development) in London in 1983, presently the largest, oldest and among the most respected NGO's outside of Africa. Dorkenoo's organization has inspired similar groups in Germany (founded in 1998), Nigeria (1999) and Somalia (2005). She has testified before the United States Congress about the magnitude of the practice of FGM and is a recipient of an award and monetary grant from the British government in light of her work, founded FORWARD in the UK to heat up the fight. Who better to tell the story of the horrors of FGM than a top notch nurse and former mid-wife who has seen the health risks, inflection and death associated with the practice? Naturally, because the guarded practice is so well entrenched in Africa and parts of the Middle East, Dorkenoo continues to take a lot of heat for her work. But she was able to reach Senator Edward Kennedy propelling him into the forefront of international advocacy against FGM. She also inspired congresswomen Patricia Schroeder and Barbara

Rose Collin's toward House Resolution 3247 against FGM in the United States. There have been many other highlights to Dorkenoo's courageous work, but the Human Rights Award she received from feminist organization Equality Now, founded in 1992 was a big deal to her. In 2005, Dorkenoo received the coveted award from hands of actress Meryl Streep at a ceremony in New York. It is Dorkenoo's book *Cutting the Rose: Female Genital Mutilation, the Practice and Its Prevention* that offers up the rose as the centerpiece symbol of the developing global movement against FGM. The rose of course symbolizes the coveted, sacred, powerful vagina and FGM involves the practices of *infibulation*, the sewing closed of the vaginal opening and *clitoridectomy*, the removal of the clitoris, both horrifying to think about. What's more, a league of Nigerian artists, including men like Godfrey Williams-Okorodus who resides in Antwerp, Belgium, have used the rose in various ways in their paintings and art work to aid in the fight against FGM. Okorodus's art work is part of an international traveling exhibition of watercolor and oil works that address the horrors of FGM.

Following the first edition of *Cutting the Rose*, Dorkenoo accomplished another pioneering feat: in July 1992, as the head of FORWARD-UK, she organized the First Study Conference on FGM in the European Diaspora which took place in London and served as the forerunner to intensive European organizing, culminating in the European Network against Harmful Traditional Practices (EuroNet FGM) founded in Brussels in 2002.

Tobe Levin, collegiate professor in the University System of Maryland in Europe lives in Germany, came to the anti-FGM movement in 1977 after reading a magazine article that galvanized her, appropriate for her profession since she teaches literature. The magazine was *EMMA*, a German feminist publication and the article was "Klitorisbeschneidung" (clitoridectomy) written by journalist Pauline Caravello. Levin couldn't believe what she was reading, "They do what?!!" was her immediate reaction. What she felt at once was empathy and rage.[289][26] Levin well remembers the day she read the article and the magazine's stillness as she stared at what she'd read with fugitive shrieks that bounced off the walls. Levin also recalls that just thinking about the contents of the article caused her to dive under her duvet with legs crossed, tight. Levin who is a wife and mother couldn't believe that so many women in countries throughout Africa and the Middle East were denied the right to make decisions about their bodies. The FGM practice is so culturally entrenched that often elder women are the chief perpetuators and promoters. Levin vowed to do something and she has for thirty years running. Immediately after reading the article in *EMMA*, Levin contacted the only African campaigners against FGM that she could find: Awa Thiam, Edna Adan Ismail, Nawal el Saadawi and Marie Assad. With an abundance of continued criticism, all of them have worked tirelessly to educate people

---

289 Tobe Levin and Augustine H. Asaah's book *Empathy and Rage: Female Genital Mutilation in African Literature*, Canterbury, England, Becky Ayebia Publisher (2009) speaks to Levin's initial outrage upon learning of the practice.

about the practice and to garner support to put a stop to it. It is seemingly lifetime work; the practice is quiet and underground except when young women end up at hospitals or clinics with inflections or worse. It has now more than seeped into the diaspora throughout Europe and the United States with hundreds and hundreds of cases ending up in the medical system. Using the work of Alice Walker and others, Levin routinely educates her students about FGM. She is a co-founder of FORWARD-Germany, companion organization to Efua Dorkenoo's group in the UK.

The women and men who fight against FGM are at all times confronted with the question: How to approach an issue viewed as urgent by only a tiny trans-national minority that is supported with tenacity by an enormous, and excessively powerful majority? It is so incredibly entrenched that as late as 2007 there is proof that the practice was actually done as a way to buy votes in Sierra Leone where 90% of females are said to be cut. One must get graphic about the horrors of the FGM practice in order to confront it. Sadly, FGM has defenders, females at that. In 1991, FORWARD board member and mid-wife Comfort Ottah confronted defenders in letter to the editor of *Emerge Magazine*, George Curry. Ottah thinks it is ridiculous to compare male circumcision with FGM and challenged notions by Harriet Washington who as quoted sought to normalize the practice in the <u>Emerge</u> article by asking if knew how many men "go out to satisfy their sexual needs because it is impossible to do so with their wives? Does she know how many women are abandoned by their husbands because they shrink away due to pain each time the husbands come near them for sexual relationship?"[290] Ottah pointed to the number of school girls who stay home from school because they cannot menstruate freely and the number who might spend 30-45 minutes trying to pass urine and are in trouble with their teachers for being late to classes? Comfort Ottah's piercing letter to the editor explained other inconveniences of FGM such as urinary tract infections and worse, a number of deaths of unborn infants is associated with the practice. Finally, Ottah compared FGM to practices like binding of women's feet, chastity belts, denying voting rights to women, slavery and so on, all efforts to oppress or control. "Enough is enough," she concluded.

A fiery voice in the anti-FGM movement is Awa Thiam, one of the women who promoted the film *Warrior Marks* in London. Born in Senegal, Thiam's best known book is *La Parole* aux *négresses* (1978). Thiam slams a number of practices that are oppressive of African women, including FGM, polygamy and skin whitening. But Thiam doesn't blame African men solely for these practices. In her book *Black Sisters, Speak Out,* Thiam reminds us that African women also keep alive practices that are oppressive of females. African men, she proffers, were pushed into brutalizing women because white colonialists had made them feel so inferior.

---

290 Ottah's letter to George Curry, published in *Emerge*, September 1996, p. 30 was taken from the exhibition catalogue, *Through the Eyes of Nigerian Artists: Confronting Female Genital Mutilation*, p. 20. Edited by Tobe Levin and initially published in 2000, Frankfurt by FORWARD-Germany.

Coda

No other flower in the universe is as associated with unusual gifts to select humans from the Source. The gifts we have witnessed in this essay run the gamut from medicine and humanitarianism to the pure genius of creating something wonderful like beautiful, timeless music or opera and literature. Apparently human beings were intended to shadow the rose and can fully do so when they step into the light of service to others or when something beautiful and beneficial is created for the benefit of all.

*with translation assistance and editorial support from Tobe Levin (Germany), Hisae Ogawa (Japan) and Alex Prodovikov (Russia)

10

"...Blossom As the Rose" in OAHSPE, *The Emerald Tablets*, and the *Holy Bible*

By Frankie Hutton

> The wilderness and the solitary place shall be
> Glad for them; and the desert shall rejoice, and
> Blossom as the rose.
>
> > *Holy Bible*, (Saint James Version)
> > Isaiah 35, Verse 1

> In obedience to the law, the word of the Master
> Grew into flower.
>
> > *The Emerald Tablets*

> By his command shall a rose bloom in our midst.
>
> > Book of Saphah, **Oahspe**

According to an expert on cultural icons and symbolism, the rose is said to hold a "royal status among flowers" explained by "its association with comfort, generosity and discretion.[291] But it is apparently not well known that high level entities in nature and in the universe have been connected to the rose. Inextricable, for instance, is the connection of the rose to the sun, to the great mystic Jesus Christ who was known as the "rose child" and to the floral symbol of the female vulva.[292] Even the magnificent scent of the flower has been revered by both pagans and patricians over epochs and across global timelines. Donald Tyson's compilation on the history of occult philosophy reminds readers that in every good matter such as love and good will, "there must be good fume, odoriferous and precious" and likewise wherever something bad or at least of no good value is brewing, there are "stinking fumes that are of no worth. "During ancient times, the essence of the rose was mixed with all sorts of concoctions including with "musk, red coral, ambergris" and even with unmentionable animal parts to create "suffumigation," according to the foundations of western occultism.[293] In modern times, the precious oil essence of rose petals has been mixed with numerous other scents to create sought after fragrances that have brought fortunes to the global perfume industry.

While the rose has been well recognized for its ostensible natural qualities and has been used as a diverse

---

291 Hans Biedermann, *Dictionary of Symbolism: Cultural Icons and the Meanings behind Them.* (New York, Penguin Books, USA, Inc. 1992), p. 290.

292 Marcell Jankovics, *Book of the Sun* translated and edited by Mario Fenyo (Wayne, NJ Hungarian Studies Publications, Inc. and Columbia University Press, 2001) p. 110.

293 Donald Tyson, editor, *Three Books of Occult Philosophy: Written by Henry Cornelius, Completely Annotated with Modern Commentary; The Foundation Book of Western Occultism* (St. Paul, MN, 2004) p. 132.

symbol globally, mystical aspects of the flower have been the least known and apparently, for most, unknowable. Nevertheless, a bevy of clues and hints about arcane aspects have been given to mankind such as the rose's enticing scent and, more than any other flower, its readiness to be crossbred. Rose petals are smooth, fragrant and could not be more perfect. A most unusual clue about the rose's special mystical place in the universe is symbolized in the fabulously colorful rose nebula mentioned earlier in the introduction of this anthology. This giant cosmic nebula spans 50 light years and lies hovering in the cosmos 4,500 light years away from earth, near the constellation Monoceros.[294] Yet few human beings have taken note of its presence probably because it can be seen only with the aid of a powerful telescope. Of the myriad shapes this provocative nebula might have taken, the rose became the resonant form in both magnificent red color and shape. It has been suggested by websites that the nebula's presence was foretold eons ago but no one knows for sure just how long this giant collection of star dust has been in the cosmos. The point of bringing the nebula to mind here is that the rose symbolically and botanically pops up throughout the universe in a variety of ways to render its message for those who have "eyes to see" so to say. The rose has quite a remarkable presence for those who have

---

294 For more on the rose nebula, see Jerry Bonnell and Robert J. Nemiroff's "Astronomy 365 Days: The Best of the Astronomy Picture of the Day" Website, (Abrams, New York, 2006). Also of interest is *Constellations, Stars and Celestial Objects*, (Firefly Books, Stuttgart, Germany, 2005)

arrived at higher awareness; it has been taken note of for hundreds and hundreds of years by those who are paying attention to one of the most important, archetypical symbols nature has ever rendered. It should be emphasized again that appearances of the rose or its likeness in the cosmos and in varied sacred literature bespeak the flower's pedagogic role. It is truly the flower of perfection and love, an apropos symbol for the global village. The mission of the rose transcends any particular religion or belief system and over time has been connected to Isis, Aphrodite, Venus, Mary, the Knights Templar, the Rosicrucians, the Order of the Golden Dawn and even to Zen.[295] Alchemists are said to have embraced it as "the flower of knowing" or "rosarium philosophorum." Its message is irrespective of religion; those who are mired in the dogma of any political or religious persuasion are the most unlikely to understand or to follow the real messages symbolized in the flower. In other words, those steeped in fundamentalist doctrinal matters so fully as to think and live separately and clannishly are blinded to the messages. The rose's messages are not for zealots of any religion; nor are the messages for the avaricious. Simply put, the rose signifies too much for those steeped in zealotry, provincialism and greed to sense or to manifest. The frequency of the rose, as rendered by nature, is higher than any particular religion or dogma, its messages apparently transcend particular belief systems and

---

295 Rudiger Dahlke, *Mandalas of the World: A Meditating and Painting Guide* (New York, Sterling Publishing Co., Inc. 2001) p. 136.

cultures as contributors to this collection have made cogently clear. This last chapter in the anthology is in no way the final word on the rose for there is much more to reflect upon and to understand as will be made clearer to each individual who comes, in time, intimately to know the flower. The evidence is that each person who strives for higher awareness has the potential to know all that the rose symbolizes as it ultimately points the way to intensification of consciousness. This is no small feat. As we have seen in this anthology, a number of extraordinary human beings have come to know the arcane life of the rose and those myriad aspects of what the flower signifies. As a companion to the last chapter about world class individuals who have embraced the impeccable flower in their work, this final chapter, the denouement of the collection, has as its focus the rose's presence in important sacred literature. Again, as in the last two chapters, we meet high level human beings connected in some way to the rose and in doing so pose the question: why is it that these people are all unusually talented leaders, larger than life so to say? Is that by design? It is it mere happenstance? Is there some organization or association that these individuals have in common? Apparently a few human beings in the universe have been pulled toward the flower and all that it signifies, in some cases after years of odd, seemingly serendipitous experiences and occurrences that have ushered them in its direction when the time was right. Sometimes initiation to an esoteric society or fraternal group provides their introduction to the rose; sometimes not. One thing is fairly certain; the flower appears after hard work, understanding and real intent toward truth seeking begins.

Knowledge and understanding of the inconspicuous presence of the flower in sacred and esoteric literature coupled with vignettes of little-known human beings who've encountered it, lead the way to insights of the rose's cosmic, mystical realms. But the real understanding of the flower comes from serious, well intended, continued work on oneself and this singular fact has been told to human beings in numerous places, veiled and unveiled. Thorns on the stem of the rose mean that some of what the flower signifies — its hidden sacred knowledge is most assuredly not intended for the profane—that is people who cannot see themselves, who do not regard others and who have not reached some measure of discipline in every aspect of their lives.

The rose is spread out in a quiet presence all over the universe, in important sometimes veiled literary passages and scriptures for those "who have eyes to see." The flower has had an overt presence in *OAHSPE*, a unique and little known nineteenth century Bible, has been mentioned in *The Secret Doctrine*, Helena Blavatasky's two-volume opus that ushered in the late nineteenth century theosophical movement, and is alluded to in *The Emerald Tablets* said to be an otherworldly document that pre-dates Christianity; and it even appears twice in some versions of the Christian *Holy Bible*. Yes, the evidence is that this quietly perfect flower has a varied presence all over the universe, in what appear to be spiritually inspired documents serving diverse cultures as far back as the lost colony of Atlantis dating thousands of years before the time mark created with the birth of Christ. The flower's veiled presence in some versions of the Christian *Holy Bible* is remarkable too, mostly because the verses containing mention of the rose in metaphor in the books "Song of Songs" and in "Isaiah" are generally misinterpreted and apparently misunderstood. [296] From diverse cultures the world over, people know at least something about the red rose: that it is lovely in the universe to touch, to smell, to enjoy and even to eat. But the masses are oblivious to veiled revelations of this ever present flower and to oblique, quiet references to it in important, urgent literature from various cultures and religions. Inimitable and timeless, the rose apparently has a cosmic connection as no other earthly flower, except perhaps its counterpart, the lotus of the East that is said to have a thousand petals and has even been referred to as the rose-lotus. The two flowers appear to be synonymous when it comes to matters of metaphysics and higher consciousness, although some groups such as the Order of the Golden Dawn use both the lotus and the rose in different ways during ceremonies. [297] To stay on point, what

---

296 The Christian *Holy Bible* has had numerous translations and versions and not all elect to use the rose in metaphor, but the most widely used King James version does mention the rose in the books of "Isaiah" and "Song of Songs." *The Holy Bible: New International Version* makes reference to the crocus instead of the rose in the book of "Isaiah," Chapter 35. Interestingly, the King James version is thought by some to have been written by Francis Bacon on behalf of King James I of England; this notion is feasible since Bacon was steeped in Masonic and metaphysical traditions and is even thought to be the true author of Shakespeare's plays and sonnets that are also full of parable and allegory that reference the divine.

297 Israel Regardie, *The Golden Dawn: A Complete Course in Practical Ceremonial Magic* (St. Paul, Minnesota, Llewllyn Publications, 1986) p. 47.

are the messages to be conveyed by the rose and for whom are the revelations intended? Well, undoubtedly, to underscore the point, the rose has been "planted" to tell humans something providential and essential from the universe-- from Cosmic Conscientiousness. In order to get the messages clearly, commercial qualities of the rose have to be put aside and this essential floral essence has to be seen as an exemplar of perfection and readiness to serve human beings in natural beauty and quiet perfection. These are precisely the qualities that human beings were intended to provide for each other and to the universe. Most human beings have clearly failed to create on earth or even to work fully toward all the good that the rose symbolizes in the universe from the smallest, most simple wild rose that is appreciated and used by Native Americans to the giant rose nebula, thousands of miles away in the cosmos.

The rose must be viewed as a profoundly symbolic link to divine aspects of the universe, as a sort of quantum theology in a flower. A select few in every generation have understood the messages and have attempted to live and even to teach the messages to those who are ready or astute enough to understand. Some human beings have been pulled toward the rose close enough to brush against it so to speak, but most never know its full splendor. Social philosopher Jacob Needleman addresses something of this phenomenon -- of never quite getting there, never quite making the connection to higher consciousness -- in his new book *Why Can't We Be Good?* "The obligation that is offered to us is to strive with all our being to serve what is good—while at the same time, also with all our being, to suffer in full consciousness the naked fact that it is beyond our strength. Then and only then can moral power be given to us."[298] As some reading this collection of essays may see, the symbolic messages of the rose are profoundly serious and are concerned as much with matters of culture, world peace and inner harmony as with individual aspiration to connect with the Divine as these goals are truly all interconnected and infused as highly evolved people come to cogently understand. One is able to glean this through connecting dots, to use a popular metaphor, and through the "understanding" and higher intuition that transcend the material world. Insights begin to come slowly after sincere work in the "right" direction. The essential, profound, "right" fact is that living in the material world with inner harmony and service to others while simultaneously aspiring to make the divine connection are simply flip sides of the same coin. Put another way, human beings are supposed to learn to work on inner harmony and at the same time to work together to connect with each other to make the earth a better place to live for all. This is truly multi-strata work that is to be undertaken daily without too much fanfare simply because it is the "right" thing to do, not because reward is expected. Of course these very messages are replete in all of the sacred literature, but have been glossed over in preference for the empty dogma of religions

---

298 Jacob Needleman, *Why Can't We Be Good?* (New York, Penguin Group, 2007) p. 252.

and material hankerings of human beings that actually mean little in the realms of seriously higher awareness. The work referred to here is really important and requires discipline to accomplish. Even those who understand what ought to be done have a difficult time doing the work consistently-- that is simultaneously disciplining and harmonizing themselves and working to make the world a better place. A classic little book in metaphysics from the eastern perspective, *The Secret of the Golden Flower,*_comments cogently on these profound, essential missions, that must be understood and done in tandem, one done without the other doesn't amount to much: "Mastery of the inner world, with relative contempt for the outer, must inevitably lead to great catastrophes. Mastery of the outer world, to the exclusion of the in inner, delivers us over to the demonic forces of the latter and keeps us barbaric despite all outward forms of culture."[299] Those paying attention can find plenty of examples of the barbaric in the world today as in centuries past. So much of what human beings do to each other and to themselves is a horror show that repeats or recycles generation after generation even among so-called highly civilized and educated people. Simply recall the lack of civility in our daily lives; some of what is happening now is surreal. Warmongering, hate crimes and domestic abuse are apparently on the rise worldwide. And because inner quiet and harmony are also indispensable

---

299 *The Secret of the Golden Flower: A Chinese Book of Life* Translated and explained by Richard Wilhelm with Commentary by Dr. C. J. Jung, (New York, Harcourt Brace Jovanovich, 1931) p. viii.

to higher consciousness, the noise and negative vibrations emitted from over-usage of cell phones, computers, iPods and, of course, high tech weapons hamper human beings from connection to the Source. Before he died, controversial writer, social activist Norman Mailer made a frontal assault on organized religion and forged a new catechism that links contemporary technology to weaponry of the devil.[300] As we witnessed in Chapter Eight of this collection, Russian Daniel Andreev writing secretively while in a gulag in the 1940s also concluded that so much of what new technology offers is an abomination that will be used ultimately to oppress and hurt unsuspecting citizens worldwide.

Big governments and religions have failed to take responsibility to bring human beings truly together. In fact, it is becoming more obvious that walls have actually been engineered to separate human beings. Too often steeped in propaganda, dogma and zealotry, religious and governmentally engineered divides have been impediments to the right and good work that must be done to raise the collective consciousness. The earth has been divided in such a fashion that immense, impenetrable walls ensure that chaos and tension remain the rule rather than the exception. There seems little hope that human beings will ever know universal harmony. We have actually been socialized and taught for thousands of years to prefer fighting and loathing to caring about each other. Basking in varying degrees of

---

300 See Norman Mailer's book, *On God*, (New York, Random House, 2007).

discord, war, hatred, and selfishness perpetuated very well by governments, religions and a few powerful individuals is a lot more profitable than building bridges to bring people to the round table of peace and concord. Those who deal in arms and weapons are having a grand old time working in tandem with a bevy of officials and their associates to make loads of money and to control, oppress and even punish citizens for no high or divine purpose other than avarice. [301] Propaganda and dogma are skillfully used by a few high leaders to maintain a status quo that is divisive, anti-democratic, self-serving and more and more counterproductive. War mongering divides have been created that go on and on, far beyond so-called lifetimes. *"My God is better than your God and I know him better than you do and will proselytize, oppress you or even kill you to prove it"* seems to be the way of the world. How dumb of human kind not to work toward and demand peace, cooperation and love. Ironically, former United States President and high ranking World War II era military officer, Dwight Eisenhower is often remembered as having said people want peace so badly that sooner or later governments will have to move over and let them have it. To Eisenhower can be added a number of world class human beings, many of whom have witnessed war firsthand and have come to realize that something is out of kilter and has been for a very long time. We seem at times now to be in a

---

301 To corroborate this notion of war, weapons and conflict at the bidding of government, see, for instance, *Blackwater USA*, 2007 that explains the development of a private military force.

slow self-destruction, ironically led by organized religion and avaricious governments backed by a few powerful individuals who will not modify their greed agendas for the good of all.

Leo Tolstoy, a Russian aristocrat and Crimean War (1853-1856) era military officer also realized the importance of peace and cooperation and wrote about these powerful ideas very convincingly in a masterful work entitled *The Kingdom of God Is Within You.* Tolstoy explained that when "war breaks out, in six months the generals have destroyed the work of twenty years of effort, of patience and of genius." Tolstoy had seen war and knew its horrors well enough to make clear his conviction that it created wretched, horrible situations and, moreover, that it resulted from the "most hideous materialism."[302] Whether Tolstoy embraced the rose symbolically or not, he seemed to know something of its real essence as did his compatriot Daniel Andreev mentioned in the last chapter of this collection. In America, Civil Rights leader Martin Luther King, Jr. also understood the rose's messages very well, although he is not ostensibly linked to the flower in any particular literature. Dr. King often used mountains as a metaphor in his speeches such as in his famous "I Have a Dream" speech. In using mountains as a metaphor, he was alluding to human beings, all, getting to a higher place where they can see the full picture: that we're all connected, all part of the same Source. We have

---

302 Leo Tolstoy, *The Kingdom of God is Within You: Christianity Not as a Mystic Religion But as a New Theory of Life,* Martin Green, editor, (Lincoln, Nebraska, University of Nebraska Press, 1984) p. 152.

witnessed in this anthology that a few people worldwide
know the rose and have come to the realization that, in
addition to its symbolic cultural connections, it signifies our
interconnectedness and potential for higher consciousness.
Still, the masses are easily led to war and they seem, as
crazy as it may be, content with divisiveness and war. Too
often religious fundamentalism becomes zealotry; people set
themselves up to think that their religion or denomination
is better than any other. With pomposity too many human
beings act as if they or their groups know God better than
anyone else; down through the ages some have been inclined
to bash heads to prove their God awareness. Inside too many
churches and denominations people are jealous, unkind and
mean spirited toward each other; outside their own groups,
it gets worse: they want to harm each other. History teaches
us that a lot of blood curdling violence has been committed
in the name of religion. Such behavior most assuredly is not
the way to know God or to reach Cosmic Consciousness that
is one and the same.

A *New York Times Magazine* article written by Mark
Lilla, professor of humanities at Columbia University in New
York City reminds us that after all, theology is nothing but "a
set of reasons people give themselves for the way things are
and the way they ought to be."[303] Too often unscrupulous,
avaricious leaders have run a show so to say that divides

---

303 Mark Lilla, "The Politics of God," *New York Times Magazine*, August 19,
    2007, p. 30.

people rather than bringing them together.[304] Mankind suffers remarkably because of immense social and psychological separations engineered by churches, political and social organizations and avaricious governmental leaders. Those who lead do not exhibit the magnanimous, kind and tolerant behavior necessary to mend tense situations. Apparently, some of the leaders do not know any better themselves; at other times it would appear that divides are purposely, willfully created. Not good.   Professor Lilla reminds us of the teachings of English philosopher Thomas Hobbes: "Messianic theology eventually breeds messianic politics"[305] thus warmongers know they can ultimately sell the masses almost anything, including rank aggression of other nations and human beings because "citizens" the world over buy the notion that redemption is always possible after the mega foolishness and killing sprees known as war. Lilla reminds us too of the wars that have resulted from religious fervor and our failure to see the real core of doctrinal literature, that we "must change our lives." We need, first and foremost, to work on ourselves to be better people at all times and in all places. This unheeded, high message is at the core of all of the major religions and metaphysical movements.   Few get the message.

Unusual, mystical qualities of the rose stretch back over centuries and in diverse geographic locations as we've

---

304 In this regard, See Jim Marr's book *The Secret War*_(New York, Random House, 2003) and Anna Politkovskaya's *Putin's Russia* (London: The Harville Press, 2004).
305 Mark Lilla, "The Politics of God," p. 50.

seen elsewhere in this collection, for instance in Monika Joshi's essay on the rose in Indian and Hindu culture and in Michael Price's chapter that is based on Native American oral history traditions which show the use of the wild rose by indigenous people as an important nutritional and food source. Italian poet Dante's acknowledgement of the flower as a sublime component of his initiation in *Divine Comedy* and Rumi's magnificent use of the rose as a metaphor for cosmic conscientiousness in his 13th century Persian poetry are other extraordinary examples in this anthology of the flower's global pervasiveness.[306] Pointing further to mysterious aspects of the rose, James Gaffarel in 1650 noted that it had even an "astral-light" body that actually outlived the life of the flower. It is not clear what prompted Gaffarel to experiment with the flower's aura, but he found something startling: when the ashes of a burnt rose were preserved and held over a lighted candle, it became a dark cloud, still in the shape of a rose, "so Faire, so Fresh, and so Perfect a one, that you would have thought it to have been as Substantial and Odoriferous a rose as grows on a rose-tree."[307] While we know nothing of Gaffarel's state of mind or level of higher awareness, his experience with the rose is only somewhat unique as he recorded it nearly 500 years ago. Others, like

---

306 Of interest is a symphony produced in Rumi's honor by Hafez Nazeri during this 800th year celebration of his birth and the United Nations declaration of 2007 as the International Year of Rumi. In September 2007, the University of Maryland's Center for Persian Studies, College Park, MD sponsored a 3-day event honoring the work of Rumi. See *The Washington Post*, August 30, 2007, p. C6.

307 As quoted in Manly P. Hall, *The Teachings of All Ages*, (New York, Penguin Group, 2003), p.347.

Gaffarel, at various times and in a variety of places globally have also had remarkable rose encounters.

More recently, the purest essential oil of the rose has been recognized as perhaps the one and only natural oil essence that can raise the electromagnetic frequency of the body. Human beings rally around 58 MHz and frequencies lower than 32 imply shutting down of the human body, death. Dr. David Stewart, a scientist, teacher and former minister in the protestant tradition may be the first in modern times to educate human beings of the power of rose oil. The author of *The Chemistry of Essential Oils Made Simple*, Dr. Stewart notes that the oil of the rose contains 11% alkanes an important hydrocarbon and other properties that render it unique in comparison to others.[308] The alkanes of the rose supposedly have properties that are not fully understood by chemists, but these contain "spiritual and healing qualities have been recognized and applied for thousands of years." Thus pure rose oil when anointed with good intent is thought to be precious and indispensable.

Perhaps even more mysterious than Gaffarel's astral-light experience with the rose or Dr. Stewart's findings about the chemical components of rose petals, is a series of paranormal episodes experienced by a physician and dentist known as Dr. John Ballou Newbrough during the last years of the nineteenth century. Newbrough was born in Ohio in 1828,

---

308 David Stewart, *The Chemistry of Essential Oils: God's Love Manifest in Molecules* (Marble Hill, MO., Care Publications, 2006) p. 153.

attended Cincinnati Medical College and advanced quickly as a result of a superior mind and training first in medicine and then in dentistry. He was also gifted in the paranormal from teenage years onward. Gaining financial independence during the California 1849 gold rush, Newbrough married the sister of a prospecting friend and settled in New York City where he practiced medicine and connected with a spiritualist group. Newbrough's clairvoyant and clairaudient powers were sharpened as he delved into contacting "out of body intelligences" but by his own admission he remained "disgusted with the low grade of intelligence displayed by them."[309] In other words, it appears that Newbrough wanted to advance faster than his spiritual guides were leading him, a fact also true of Elizabeth Haich who admits in her book Initiation that she begged for help of a master to be advanced cosmic consciousness.[310] Apparently some human beings are chosen for initiation to higher realms; others seek initiation but are denied. Why this is the case is not easily understood but it can be deduced through the study of metaphysics that past life karma, akashic records, and sincere heartfelt intent all have a bearing on illumination. So too do hard work, service, silence and discipline of one's body, including diet. Nevertheless, between 1871 and 1881, while living in New York City in the vicinity of what is now Pennsylvania Station, Dr. Newbrough is said to have received "spiritual

---

309 John Ballou Newbrough, *OAHSPE, A New Bible in the Words of Jehovih and His Angel Ambassadors* (London, England, Kosmon Press, 1942) p. iv.
310 Read Elisabeth Haich's account in *Initiation* (Santa Fe, NM, Aurora Press, 2000).

guidance" toward purification to the extent that he was told to stop eating meat, to discipline himself and to purchase a typewriter, an instrument that had just been introduced to the marketplace, which he did. As the little known Newbrough story goes:

> Upon sitting at the instrument an hour before dawn he discovered that his hands typed without his conscious control. In fact he was not aware of what his hands typed unless he read what was being printed. He was told that he was to write a book - but must not read what he was writing until it was completed. At the end of a year when the manuscript was completed he was instructed to read and publish the book titled OAHSPE, a new Bible.[311]

With guided hands, what Dr. Newbrough produced from that new typewriter was a strange, absolutely remarkable bible known by a select few as OAHSPE. It consisted of 36 books, over 800 pages, a number of illustrations, pictographs and was a provocative rendering to say the least. One of the books or chapters in OAHSPE known as "Cosmogony and Prophecy" tells "what light, heat, electricity, gravity, etc. [are] and what causes them, what holds planets in their places, gives the many cycles of time used by the ancients in their tables of

311 OAHSPE, p. v.

prophecy and tells of relativity" and more. Another segment, The *Book of Jehovih's Kingdom on Earth* is cogent in its instructions to human beings who want to "develop spiritual powers, prophetic abilities and extra sensory perception." One warning is particularly clear and cogent: humans ought to stop eating blood thirsty carnivorous animals and stick to "herbivorous foods."[312] Prophetically, the *Book of Jehovi's Kingdom on Earth* warns of "a disease [that] came among the cows and the physicians forbade the babes being fed on their milk." Corn and rice milk were suggested as "an excellent liquid food for infants."[313] Otherwise, the same book admonishes humans to know the importance of learning: "And ye know that all light is progressive. Ye cannot settle down, saying I know enough." *The Book of Discipline* provides general instructions as to how humans are to treat each other: *very well*. Discipline in all matters is a key aspect of quantum connection. A verse in the *Book of Discipline* is cogent in its message that connection to cosmic conscientiousness is possible by abandoning "earthly habits and desires" and "by constantly putting away the conditions below."[314]

A lot of what is written in OAHSPE makes good sense. Some of it is so other worldly as to be indecipherable. Remarkably a number of chapters of OAHSPE contain short passages using the metaphor of the rose; there are even

---

312 OAHSPE, *Book of Jehovih's Kingdom on Earth*, Chapter VIII, Verse 8, p. 814.
313 OAHSPE, *Book of Jehovih's Kingdom ...*, Chapter XI, Verse 29, p. 818.
314 OAHSPE, *Book of Discipline*, Chapter III, Verse 14, p. 837.

pictographs of a rose in some books, most notably in the *Book of Saphah* that contains the Tablet of Hy'yi:

> Behold, the rose, deep rooted in the earth, Jehovih riseth in majesty of All Light. His colors no man maketh, nor knowth any man the cause. This subtle Perfume, whence cometh it, and whither goeth it? What power fashioneth it, And propelleth it?[315]

The message that the rose is rooted deep in the earth has multiple meanings, two of which are inherent in this essay. The first and most important prong of the message alludes to the human potential to open unseen chakras or energy centers of the body.[316] These chakras open like a flower when a person has done continual sincere work on themselves, service to others and has overwhelming pure intent to reach Cosmic Consciousness or what is known commonly in the Protestant world as Christ Consciousness. People who have accomplished some level of work seemingly have a glow or radiance about them that is sometimes noticed by others but especially comes to the attention of those who have worked on higher awareness. Unevolved humans in the presence of these special "light" beings simply note that they are "different."

315 OAHSPE, *Book of Saphah*, p. 618.
316 See C.W. Leadbeater's *The Chakras* first published in 1927 by Quest Books, Wheaton, Illinois and now in its 9th printing. On page 7, Leadbeater notes that there are seven chakras. Also of interest is Ruth White's *Using Your Chakras: A New Approach to Healing Your Life* (New York: Barnes and Noble, 1998).

What's more another distinguishing characteristic is that such individuals avoid the fray and foolishness of material life. We have also taken note in several chapters of this collection that the rose has important medicinal value and that its scent is invaluable. In OAHSPE, the pictorial referred to as the Tablet of Hy'yi in the *Book of Saphah* is perhaps decipherable for those who have had consistent study of metaphysics or theosophy. The interpretation given here is quite likely the first ever to be given in print. Page 618, plate 81 of OAHSPE contains a very large rose at the bottom of a pictograph and at the top of the same illustration is a crown with a cross at the top. In rectangular shape, the pictograph has two hearts just at the top of the rose. A circle with a single straight line drawn through is nestled between the two hearts. The same circle symbol is also seen in Helena P. Blavatsky's *Secret Doctrine* and is explained there in Volume II on page 30 as in connection to the root-race of the "sweat-born." Above the circle symbol is an eye and a diamond that in all probability signifies the light or ultimate perfection that a human being comes to know through the practice of pure unconditional love on earth that is essential to rise up, incorporeally, to celestial heights. Rising up is indicated in the pictograph by what is above the moon and star in the pictograph: a crown that is sure to signify cosmic conscientiousness. Not tangential to the pictograph is the fact that a perfect diamond is referred to by the diamond industry as a paragon and contains 100 carats or more, obviously when this state is reached the rose is fully opened because it has found light. In other words, we can deduce that when a human being works really hard to overcome the material world and to love and to serve other humans and to develop a sense of community, that being may potentially rise up toward a connection with cosmic consciousness or what has been termed by some as Christ Consciousness. In Volume II (*Anthropogensis*) of *The Secret Doctrine,* HPB explains that humans have the potential to drop the external shape surrounding them or to cease to be of the material world—that is to become absolutely Divine. There is enough evidence that this process can and does take place on earth as in the obvious example of the Holy Spirit mysteries connected with lives of Jesus Christ, Mirtha and others.   There are among us masters yet and still, but their presence can be sensed only by those in true service and deeply committed to the work that must be done. In this regard, today a number of groups such as Eckincar and even governments are interested in the process of rising up or detaching from one's own body, also controversially known as astral travel or as those in the world of military internationally call it, remote viewing.[317] While it has been proven that some evolved human beings actually have this capability, astral travel does not necessarily infer illumination or full initiation into Cosmic Consciousness, the so-called eternal life. Much has been written about the process of rising

---

[317] Regarding remote viewing, of interest are Paul H. Smith's *Reading The Enemy's Mind: Inside Star Gate* (New York: Published by Tom Doherty Associates, 2005) and Jim Schnabel's *Remote Viewing: The Secret History of America's Psychic Spies* (New York, Bantam, Doubleday Dell Publishing, 1997).

up to the realm of the masters; but extremely few are invited into this region. By all accounts, to move in this direction requires extraordinary discipline, inner harmony and good intent toward others. Discord and lack of good works are sure ways not to be connected to higher consciousness.

A number of verses in <u>OAHSPE</u> mention the rose briefly in a simple connection to the strive for perfection. The *Book of Apollo*, for instance, beseeches readers to "Behold the rose and the lily; they are perfect in their order. Being one with Jehovih, they painted not themselves." This book advises "symmetry of flesh; the symmetry of spirit; the harmony of music, [and that humans should] "consider wisely their behavior." Once again, this message is contained in practically all sacred literature although it is one of the most unheeded revelations in the universe.

A paradoxical footnote to the work of Dr. Newbrough is his attempt to establish a utopian community known as Shalam near Las Cruces, New Mexico in 1884 in order to teach others to practice many of the precepts of perfection outlined in *Oasphe*. All-encompassing for Newbrough and his friend Andrew Howland, Shalam was one of many utopian communities begun in America in the 19th century. Howland and Newbrough put their life savings into the community dedicated to training a diversity of orphaned children to be disciplined and spiritually exceptional. The successes were remarkable as described by Lee Priestly who has researched the Shalam community:

Naturally the spiritual development of the children was of prime importance. In addition to the study of OAHSPE, they were taught the means of communicating with angels and how to discriminate between good and evil spirits. At the ages of twelve to fourteen they were initiated into the rites and ceremonies of the Ancients with explanation of appropriate signs, symbols and emblems. Trances and manifestations were common-place to them. They received spirit messages and responded to rappings and table tippings.[318]

According to Priestly's research, the children were also taught manners and decorum. The children were the focal point of the community and were trained to be unfailingly polite, but not forward. They conversed with adults easily and with poise; among themselves they were loving and noncompetitive.[319] Unfortunately, the same cannot be said of the adults of the community known as Faithists. Not surprisingly, the adult group at Shalam was described by Priestly as men and women who shirked labor, expected something for nothing, lounged idly and were known generally to be quarrelsome and critical. The type of behavior exhibited by adults at Shalam seems to be typical of human beings who

318 Lee Priestly, *SHALAM; Utopia on the Rio Grande, 1881-1907*, (El Paso, Texas Western Press, 1988), p. 32.
319 Ibid., p. 33.

have not come to understand the work on themselves that is absolutely necessary for higher awareness.

At age 63 in April 1891 Newbrough and a number of the children at Shalam died of la grippe, a particularly virulent disease so named because of the horrible grip it had on some American communities at the end of the century. With Howland at the helm, the Shalam community lumbered along for over a decade after Newbrough's death but was beset with law suits and bickering among the adults who lived there. Howland was said to be fair and reasonable and he introduced many new vegetables and fruits to the area, but was not a manager of the caliber Newbrough had been. Noted for his strange attire of white pantaloons and preferring a long flowing white beard, Howland was arrested twice for "indecent exposure." [320] Shalam folded in 1907.

Perhaps stranger than fiction or mere coincidence, Helena Petrovna Blavatasky, also telepathic and one of the founders of the worldwide Theosophical movement, died the same year as Dr. Newbrough in 1891. As noted earlier, Newbrough was born in Ohio in 1828 while his contemporary, Blavatsky, was born in Russia in 1831. There is no evidence that the two knew each other. They traveled in different circles, he in the United States and Australia and she in India, the United States and all over Central Europe where the *Secret Doctrine* was penned. The evidence is that the *Secret Doctrine*, well over 2600 pages, was actually

320 Ibid., p. 22.

spiritually dictated to Blavatsky while she resided in several locations including in Elberfeld, Germany, Ostend, Belgium and around England with Countess Constance Wachtmeister as her companion a good portion of the time. The two were frequently in the company of visitors from the Theosophical Society who were witnesses to some of the unusual circumstances surrounding Blavatsky's work on the opus.[321] By HPB's own account she was visited by an ascended master in 1851 while visiting Hyde Park in London with her father for the first time; she was quite young at the time, only 17 years old.[322] It is generally thought that HPB's guide was Master Morya who is said to be the head of "all of the esoteric schools which truly prepare an aspirant for ashramic contact and work."[323] But long before 1851 there was a lot of drama and mystery surrounding the privileged Russian woman known as HPB among theosophists. To add to her mystique, she is known to have run away from a seventy-five year old retired military husband at age 17 after declaring herself a widow who "wouldn't be a slave to God Himself, let alone man."[324] Known to have a hot temper, Blavatsky was undoubtedly an oddity by all accounts. Those who came into her orbit were likely to witness some very

---

321 Countess Constance Wachtmeister et al., *Reminiscences of H. P. Blavatsky and the Secret Doctrine*, Wheaton Illinois, Theosophical Publishing House, 1976.

322 Master DK or Djwhal Khul is said to have been HPB's guide or master. Of interest regarding ascended masters is Phillip Lindsay's book, *Masters of the Seven Rays: Their Past Lives and Reappearance* (Queensland, Australia, Apollo Publishing, 2006), p. 49.

323 Ibid., p. 41.

324 Mary K. Neff, editor, *Personal Memoirs of H. P. Blavatasky* (New York, E. P. Dutton & Co., Inc. 1937), p. 37.

strange, unexplainable things wherever she was around; by all accounts she was much stranger than her contemporary Dr. Newbrough. Because she was so remarkably unusual, stories about her are kept alive because a number of people were aware of her "gifts" and tracked her from an early age. Key among those tracking HPB were members of her family. Known as Helene von Hahn at birth, HPB was recognized as a medium from her childhood years and known for very high intellect in the years after her mother's death in 1842. She was also "the strangest girl one has ever seen, one with a distinct dual nature...one mischievous, combative, and obstinate— everyway graceless; the other as mystical, metaphysically inclined. ... No schoolboy was ever more uncontrollable or full of the most unimaginable pranks..."[325] Manly Hall's insightful article about her in *The Phoenix* benefited from the journal notes of HPB's aunt in Russia:

> It was as though a troop of sprites were in constant attendance, ever mindful of her bidding. Her uncanny prophetic powers by turns amazed and discomfited visitors for looking them intently in the face, like some Pythoness or Delphi, in half-formed childish words she would solemnly predict the place and time of their death. .. [H]er prognostications were so correct that she became the terror of the domestic circle. Nothing could

325 Ibid., p. 26.

concealed from her that she desired to know. She could read the most hidden thoughts and motives and was constantly aware of circumstances occurring at a distance.[326]

Dr. Newbrough, it may be recalled, was also approached psychically in a preliminary visit from a master, but much later in his life than HPB and closer to the time he was actually dictated his one and only opus, OAHSPE. There is no evidence that either HPB or Newbrough was fearful or apprehensive about undertaking the tasks given them by masters. While there are some similar aspects in the way the two were approached, there is no evidence that Blavatsky was asked to be careful of her diet or to purchase a typewriter as Newbrough had been instructed. Other than both being "visited" preliminarily years in advance of the major tombs they were guided to write, the only other noteworthy similarity is that both Newbrough and HPB were plagued by scandal and charged by "outsiders" as being impostors. It could be reasoned that Blavatsky's reputation suffered most from scandal since her work came to the attention of theosophists, non-theosophists and even the global press followed her work rather closely. The attempts to discredit HBP are not surprising since she was guided by an other-worldly source that few human beings can comprehend then or now. As Manly Hall explained in an early issue of *The*

---

326 Manly Hall, editor, *The Phoenix: An Illustrated Review of Occultism and Philosophy* (Los Angeles, Hall Publishing, Co., 1931-32 edition), p. 86.

Phoenix, "humanity attacks viciously and relentlessly anyone who assails the infallibility of the mediocre." What's more, "The fear is not that the occultist may be wrong; the fear is that the occultist may be right."

Ironically, the information dictated to Blavatasky contained two major divisions known as *Cosmogenesis* (Volume I) and *Anthropogenesis* (Volume II) which are also aspects of the voluminous revelations dictated to Newbrough. Writing, terminology and organization of the two books are vastly different and *OAHSPE* contains far more illustrations than *Secret Doctrine* but both are clearly major works that have been overlooked by the masses. It is pure conjecture to say that both works have been reserved for or intended for a select few, but it would appear that is the case. Is it mere coincidence that the symbol of the rose appears quietly in both *OAHSPE* and in *The Secret Doctrine* or that the two authors died the same year after having been given major Arcanum or keys to understanding the cosmological foundations of the universe? What Newbrough and Blavatasky provided apparently far exceeds the cosmological and anthropological insights presented in any other religious or sacred literature, yet few are aware of their respective works, *OAHSPE* and *The Secret Doctrine.* Ironically, HPB and Newbough were only three years apart in age; she died at age 60 and he at 63. Of course there is now and always has been a great deal of conjecture and speculation about the so-called lost continent of Atlantis; but highly astute, great minds such as Plato and Helena Petrovna Blavatasky say firmly that the continent did indeed exist. Whatever one thinks of HPB's pronouncements about Atlantis or the oddities surrounding her life, she was undeniably a great intuitive and was recognized widely as such. Blavatasky explained that "axial disturbances" are part and parcel of the "intelligent Kosmic hand and law" which alone could reasonably explain such sudden changes..." [as the disappearance of Atlantis]. According to HPB's *Secret Doctrine*: "Old continents were sucked in by the oceans, other lands appeared and huge mountain chains arose where there had been none before." What is the evidence of all of this? Plato's *Timaeus* also discusses Atlantis in some detail that has been pretty well corroborated for those who have taken note.[327] And HPB offers, as evidence, the very large statues found in the Easter Islands and in the Gobi Desert indicating that Atlantis did exist. Human-like beings over "nine yatis" or "27 feet tall" inhabited Atlantis which was the Atlantic Ocean portion of Lemuria; this was the cradle of the third root race that occupied a vast area of the Pacific and Indian oceans, according to HPB.[328] These are apparently the same large statues depicted to the world in the Erich von Daniken controversial film "Chariots of the Gods" nearly 35 years ago.[329] If we can concede for the purposes of this essay

---

327 Noteworthy is Lewis Spence's *The History of Atlantis* originally published in 1926 by Rider & Son, London and New York. Plato's work, *Timaeus* was published 400 years before the birth of Christ.
328 See H. P. Blavatsky's *The Secret Doctrine: The Synthesis of Science, Religion and Philosophy, Volume II, Anthropogenesis* (London, Theosophical Publishing Company, 1888) p. 333.
329 See Erich von Daniken's film *Chariots of the Gods* released in 1972 by Sun Classic Pictures and packaged in 2005 by Blair and Associates, Ltd. Von Daniken's controversial theory is that man did not descend from apes as maintained by Charles Darwin, but that humans have descended from gods.

that Atlantis did exist, it would make sense that the giant beings who lived there were pretty smart to leave statues reminiscent of themselves made of a rock substance that has lasted thousands and thousands of years. And it wouldn't be surprising that at least a portion of this Atlantean group was so highly evolved that they lived by a set of laws set down in a special, little-known book called *The Emerald Tablets*. The leader of this evolved group of Atlanteans in a place known as Keor was Thoth, an eternal being who knew the secrets of everlasting life and led a group of thirteen adepts who were said to be well versed in scientific and philosophical matters.[330]

In recent times, there has been renewed interest in *The Emerald Tablets* and conjecture whirls about the origin of this obviously important work, including who fetched it back to the Pyramids in Egypt. Michael Doreal is said to be the man who revived and translated the *Tablets*. Doreal was a great if not well known spiritual leader, healer and teacher who was born Claude Dodgin in Oklahoma (USA).[331] Rather amazingly, he was thought by some to be a reincarnation of the great teacher Horlet who lived in Egypt during the time of Atlantis, at least 20,000 years ago. Although there is no evidence that the occurrences of his life were as controversial or as stupefying as those connected to Newbrough or HPB, there were some spiritual oddities, most notably the fact that

330 Michael Doreal, *An Interpretation of the Emerald Tablets Together With The Two Extra Tablets* (Castle Rock, CO. Brotherhood of the White Temple) p. 8
331 Whitby, Delores, *Doreal: As I Knew Him* (Sedalia, CO., Little Temple Library, 1980) p. 5.

he was "given" the task of retrieving the mysterious *Emerald Tablets* from the jungles of the Yucatan in a Temple of the Sun God and returning them to the Great Pyramid in Egypt. Through great peril, Doreal is said to have translated and returned the sacred, secret *Tablets* to the designated place; under his direction the book was published in English in 1939. Doreal also taught higher awareness, metaphysics courses for a number of years to a select group of males and females in Oklahoma.[332] On earth, Doreal is best known by a few as the leader of the American wing of a group known as the Temple of the White Brotherhood that has its foundation high in the Himalayas of Asia. Remarkably, *The Emerald Tablets* contains numerous mentions of a rose-like flower that has petals. By all indications, the reference is to a rose but it doesn't really matter if the flower referred to in the *Tablets* is a perfect red rose or not; the messages are cogent, clear and very much like messages in other sacred literature regarding an impeccable, revered flower. Planted in the human body are invisible flowers along a high level invisible "yellow brick road" to cosmic consciousness. The trick is to uncover the map to the path and to symbolically make roses bloom along the way through service, love and inner harmony. Notwithstanding the trials and tribulations of life, ultimately the opportunity exists to walk into the tremendous light at the end of the journey. This and only this is the essential veiled message of numerous classic stories, including *The*

332 Ibid., p. 15.

Wizard of Oz. It is also the ultimate message of the classic opera *The Magic Flute* and revered literature engineered to reach a diversity of the masses. Only a select few are drawn to this literature and yet, the messages have been prevalent in a variety of writing and pictographs through the centuries and longer. *The Emerald Tablets*, eons old and not surely not easily understood, have much to say about this process of making flowers bloom: "In freeing the consciousness from the body it is best to expand the solar-plexus (one of the chakras), the Flower of Life of the body, and send the life force flooding through it so that the body is vitalized in preparation for the consciousness to leave..."[333] Of course, it should be noted again that this work of making roses bloom in the human body requires intensive, well intentioned work that must be accomplished in silence. *The Emerald Tablets* underscores an important fact about higher consciousness: "He who talks does not know; he who knows does not talk." For the highest knowledge is "unutterable."

It is not surprising that the rose also makes its appearance quietly in the King James version of the Christian *Holy Bible* and also in Gnostic portions of the Bible such as the *Book of Enoch* that have been omitted, for whatever reason, from the King James version. Thought to predate the New Testament, *The Book of Enoch*, provides powerful rose metaphors such as the birth of Methuselah's strange son whose "body was white as snow and red as the blooming

333 *The Emerald Tablets*, p. 132.

of a rose...[334]" Undoubtedly this metaphor references Methuselah's illumination or connection to the Holy Spirit. By now it is apparent as a result of recent discoveries and insight that the Bible has been altered numerous times[335] and a significant part, the Gnostic portion, has been hidden from the masses. Still the allegories, parables and metaphors contained in this revered book, if truly understood, are instructive and, to say the least, full of revelations. Quite apparently there is no other book like it on earth although it clearly is not based on historical truth or intended for literal consumption. In other words, reading the Bible and quoting from it does not automatically translate into understanding the messages contained therein. One of the beauties and enigmas of the *Holy Bible* is that it is full of dual meanings. Moreover the *Holy Bible* contains hidden knowledge, in parable, allegory and metaphor that simply is not intended for those who are not ready for it. And a lot of it has been cut out or omitted for good reason. Geoffrey Hodson, well known as a teacher and leader in theosophy explained carefully in *The Hidden Wisdom in the Holy Bible*, Volume I that there is much support for a symbolic reading of the *Bible* in, for instance,

334 R.A. Charles, editor, *The Book of Enoch* (San Diego, CA, The Book Tree 1917, 2006) p. 91. Another reference appears on page 64.

335 The Bible has been altered and translated a number of times. That this is a fact has been noted by a number of scholars. In *The Teachings of All Ages*, page 543, for example, Manly P. Hall explains that regarding the King James version, Sir Francis Bacon was the translator given the task of checking, editing and revising the Scriptures. According to Hall, the very first edition of the King James version contains a cryptic Baconian headpiece. Interestingly, Phillip Lindsay in *Masters of the Seven Rays* notes that Francis Bacon "wrote secret ciphers ...for the Holy Bible [King James version] and [for] Shakespearean plays; see p. 29.

"the promises of perpetual prosperity and divine protection made by God to Abraham and his successors with subsequent defeats by invaders, exile under their commands in Babylon and Egypt and destruction of the Temples of King Solomon" etc. More recently, Episcopal Bishop John Selby Spong, author of *Rescuing the Bible from Fundamentalism* , *Born of a Woman* and other significant works has made clear his view that much of the Bible is misunderstood and misinterpreted by theologians. That said, it should be noted there are only two somewhat obscure, but absolutely beautiful references to the rose in the Old Testament: in "Song of Songs," chapter 2 and in "Isaiah," chapter 35, verses that are as easy to miss as they are misinterpreted and widely disputed.

"Isaiah" is said to be the most prophetic book of the Old Testament, revealing unique prophecies regarding Immanuel and the "Suffering Servant." The prophet Isaiah is also said to have been a literary genius. Of all the prophets he is said to have looked further into the future than any other as is revealed in this book of the Old Testament. According to some sources, he lived a long time, through the reign of three kings, and was even an advisor to one of them, Hezekiah, from 729 to 699 B. C.

The 35th Chapter of Isaiah, verse one uses the rose in metaphor. It reads "the desert shall rejoice, and blossom as the rose." In translation, the metaphor means after the necessary work has been completed, the tribulation period will be over and blessings will flow. This verse has nothing to do with a desert per se and this is obvious to the astute reader.

One interpretation puts it another way: "all spiritual evil and physical catastrophes will be reversed and the land and people will be blessed." In fact, the book of Isaiah is replete with this same message in veil: we all have the potential to connect with the Holy Spirit, if we reverse our courses, slay the ego, care for one another, and work fully toward the light. When this is accomplished even the "tongue of the dumb will sing" to use another metaphor for rising up out of the material realm which brings everlasting joy and no "sorrow and sighing."

A bevy of Biblical scholars have offered interpretations of the "Song of Songs" considering it allegorical, typological, as an anthology of love songs, as a three-character interpretation and as a literal love story, but the overall purpose is said "to show the joy of married love as a gift of a good and loving God."[336] All of the interpretations seem plausible, but from this writer's perspective none gets at the real meaning, the hidden message of this most beautifully written book of the Old Testament. Like countless other dual meaning verses in the Bible, verses about the rose have multiple meanings. Apparently this was the intended result because, like the sages and mystics before him, Jesus taught in parables and the Bible

---

336 *Holy Bible King James Study Bible* (Nashville: Thomas Nelson Publishers, 1988) 1010. The interpretations from this version are divided thus: (1) allegorical: treats the view of early Jewish literature, denies the literal aspects (2) Typological: in this view marriage is seen as a type of Christ and the Church (3) Collection of love songs: says the Song of Solomon is just love songs with no unified meaning (4) Love Triangle: that song of Solomon is about the love triangle or "eternal triangle" with Solomon as the villain who tried to lure the maiden from her shepherd-boyfriend and (5) Literal love story: view that this book is just that, a love saga.

is fully allegorical and metaphoric. Clearly, contained therein are sacred, arcane messages intended for a select few; one has only to look at the murder, mayhem, avarice and lack of humility in the world today to understand why mystics were careful not to waste time on "gross wits" and on those who refused to do the necessary work.[337]

At a glance "Song of Songs" has been interpreted to be erotic verse or even a love song sung about two lovers; the shepherd-king and his beloved outcast maiden. The story goes that after a period of absence, the former shepherd Solomon returns as King and takes the maiden away with him in a royal coach to become his bride. This quiet verse is reminiscent of Rumi's poetry. The essential point to be made is found in the way Solomon and his maiden embrace each other, in verse 13 of chapter one, for instance, she wants him "betwixt her breasts." But there is nothing carnal in her wishes. The metaphor between her breasts means transcending this world. Getting beyond duality that so drives the nature of human beings, or for those who've studied the "Law of the Three in Sacred Geometry," it is getting beyond the polarity of the material world. In the tradition of Taoist teachings, getting between the maiden's breasts would be tantamount

---

337 Hidden knowledge and the secret teachings of Jesus are mentioned and alluded to throughout the *Bible*, including in Mark 4:11 which says "Unto you it is given to know the Kingdom of God: but unto them that are without, all these things are done in parables." That is to say, many people worldwide read the *Bible*, but very few fully understand it. It is a tremendously complex book that takes years of study, reflection, and sincere intent to understand. This is also made clear in the Book of Mark, Chapter 4, verse 12: "Seeing they may see and not perceive; and hearing they may hear and not understand; lest at any time they should be converted and their sins should be forgiven them."

to getting beyond the yin and the yang or the dual nature of man to transcend the material world. As human beings move toward objective conscience, higher and higher levels can be achieved, usually over lifetimes up to merger with the Divine, far beyond the duality that so plagues humans on a daily basis. In the book "Song of Songs," if the King can get between, indeed beyond the maiden's breasts, he will get better than all the "other daughters" can possibly render unto him meaning metaphorically of course that the King will simply have overcome the material world. Here there is nothing erotic intended though at a glance it would seem so. In Chapter 2, verses one, two and three using the rose as a most powerful metaphor the maiden steadfastly declares that she is the best:

> I am the rose of Sharon and the lily of the valleys, so is my love among the daughters as the apple tree among the trees of the wood, so is my beloved among sons. I sat down under his shadow with great delight, and his fruit was sweet to my taste.

If this most beautiful verse of the Old Testament was put in the vernacular of today, the translation would be that the maiden referred to in "Song of Songs" is "a knock out," she's so well built, so fantastic, has it "going on" to the point that none of the other "daughters" or females in the community measure up to her. Such is the case metaphorically when one reaches Cosmic Consciousness, nothing else measures up; after the connection is made everything is magical and perfect. Nothing else is desired or needed! One literally and figuratively rises up above this material world, above this earth to connect with the source. What Dante, Rumi, Hanai, the prophet Isaiah and King Solomon all seem to know, few of us are capable grasping and certainly are not capable of doing the disciplined work to realize: there is absolute perfection if and when one ratchets up to the Holy Spirit. There is no denying that the Spirit is the key and it takes work to connect with it.[338] Sadly millions upon millions attend mosques, churches, masses and perform other religious rituals daily and have not come to understand this profound fact. The kind of attendance needed is attendance to oneself to be a better human being and to connect with respect and kindness to each other, not necessarily attendance at any particular service, religion or sect.

Considering the extraordinary occurrences and facts unfolded in the essays of this collection, all is still not said and done about the rose; there are yet other odd components of this most quintessential flower. Perhaps no aspect of the flower's symbolism is more striking and perhaps confusing than its connection to Parseeism or Perseism, the worship of defied fire. This is not fire in a literal sense, according

---

338 This knowledge is replete in the Christian *Holy Bible*, sometimes in allegory, sometimes clearly stated such as in Mark 1:4 "I baptize you with water, but he will baptize you with the Holy Spirit" and in Mark 3:28, "... But whoever blasphemes against the Holy Spirit will never be forgiven; he is guilty of an eternal sin." In fact, there are over 60 references to the Holy Spirit or Holy Ghost in the King James version of the *Holy Bible*. One can only guess at the number of times it is mentioned in numerous other versions.

to Hargrave Jennings, it is "the inexpressible something of which real fire or rather its flower" is symbolized in the rose. Fire worship is an odd belief system and philosophy that rarely even makes its way to philosophy textbooks today and yet it has been a component of pagan and primitive worship for thousands of years and continues in some traditions even today. As strange as it may seem, this connection to fire is one of the early justifications for the practice of cremation which has been around for thousands of years.[339] Even the torches used at early funerals were not solely for light. Torches, light, candles and so on when used in sacred services are of course the "inexpressible mystery of the Holy Ghost."

Fire has had a strange history in religion and metaphysics. The realities of fire's connection to the realms of religion, metaphysics and God seeking in general have actually been stranger than fiction could be, ever. The world over, for thousands of years, people haven't known quite how to make their connections to God or cosmic consciousnesses so that they make up stuff as they go along, some of it so macabre as to be mind-numbing. Some humans, over epochs, have quite apparently had a huge crazy misunderstanding about the meaning of fire in connection to religion and God-seeking. Centuries ago, and perhaps even now in some places in the universe, people actually burned animals, themselves and even each other as sacrificial offerings in attempts to

---

339 Beverly Moon, editor, *The Encyclopedia of Archetypical Symbolism: The Archive for Research in Archetypical Symbolism*, London, Shambhala, 1991) p. 308.

curry favor with God or at least their gods.[340] At times self-immolation or sacrifice has been practiced in Indian and Tibetan Buddhism and in some sects of the Middle East, the whole purpose of which was thought to be a way of leaving the material body. All of the sacrificial rituals and goings on have been most assuredly unnecessary and a perversion of what was intended. While fire symbolically relates to Spirit, undeniably, it is not to be taken literally; self-immolation profits one nothing. The point of the best theology, too often veiled, is to learn to live together in harmony and also to find inner peace. When these two goals can be accomplished in tandem by individuals, a light comes on and gets brighter and brighter—that is to say an internal, invisible flower opens and the individual potentially becomes one with Spirit.

The *Holy Bible* (St. James version), has numerous metaphors and allegories related to fire, most notably in the book of Daniel: "a fiery stream issued and came forth before him…" from this biblical verse we note that the manifestation of God was fire or Spirit. And the book of Matthew warns that man baptizes with water, but eventually "He shall baptize you with the power and glory and with fire." What kind of God would appear as fire and also promise to baptize humans as fire and not with water? For the fire philosophers, the answer to that question was easy: a God who is one with the Holy Spirit and is the Source of All and Everything.

---

340 See Marcell Jankovich, *Book of the Sun*, p. 120. Also of interest is Mel Gibson's movie, "*Apocalypto*" (2006) that graphically depicts the practice of human sacrifice during the Aztec empire of the 15th century.

As Michael Doreal explained in his interpretation of *The Emerald Tablets*, "Man's destiny is the final blending with light even though he moves through darkness during material incarnations." [341] This is the essence of why some cultures have preferred cremation and why fire has been a component of burial ceremonies. It is also sadly, for the misguided, the reason for self- immolation. Thus the "signification of fire burial is the commitment of human mortality into the last of all matter, overleaping the intermediate states; or delivering over of the man-unit into the Flame –Soul, past all intervening spheres or stages of the purgatorial.[342] Surely fire symbolizes the Holy Spirit and, as we have seen throughout this anthology, the rose symbolizes much but nothing so urgent and important as its connection to fire that is symbolic of the Divine. When the magnificent rose opens to the light, or put another way when petals of all of the invisible chakras in the human body open to the light of what is good and harmonious this symbolizes that connection with the Holy Spirit or Cosmic Consciousness has been achieved.

The great philosopher, alchemist, physician Philip Theosophratus Aureolous Bombast Hohenheim, widely and controversially known by the name Paracelsus (1493- 1541) understood the connection of fire to the Holy Spirit as is revealed in his writings and teachings. To say the least, Paracelsus did not live a charmed life. But he did live such an exciting and unusual life that stories have been written about

---

341 Michael Doreal, *An Interpretation of the Emerald Tablets,* p. 49.
342 Hargrove Jennings, p. 111.

him, including one by Argentine writer Jorge Luis Borges (1889-1986) titled "The Rose of Paracelsus." In the Borges' story a doctor (Paracelsus) has asked God for a disciple and one appears who is apparently full of doubts because he asked for evidence so that he would know for certain if he ought to follow the great doctor. In the story Paracelsus listens intently to the disciple's request for evidence—that is a request that he turn a rose into ashes and make it reappear which the great doctor declines to do. Alas, the would-be disciple leaves unconvinced that he should follow Paracelsus because he did not see the rose converted to ashes. After the disciple exits, Paracelsus quietly makes the rose appear from the ashes. The moral of the story seems to be that if humans doubt the power of the rose, the power of the Holy Spirit, then it's not for them and the connection cannot be made. Known as an efficient astrologist, alchemist and philosopher, Paracelsus was also a preacher who addressed crowds in taverns in a mix of religion and political polemic that called for social equality based on Christian principles.[343] In words that are as apropos today as they were during the Renaissance years in which he lived, Paracelsus warned, "No good can happen to the poor with the rich being what they are. They are bound together as with a chain. Learn you rich, to respect these chains. If you break your link, you will be cast aside."[344] We cannot deduce from the teachings of Paracelsus that he necessarily

---

343 Philip Ball, *The Devil's Doctor: Paracelsus and the World of Renaissance Magic and Science* (Farrar, Straus and Giroux, 2006), p. 124.
344 Ibid., p. 124.

favored the poor over the rich, but what he advocated was cooperation between the strata of society and between the races, a melding of aims for the benefit of all. Ultimately, finally, such cooperation follows the spirit of the rose. It is wise for us to mimic this spirit and to be ever mindful of the Source.

Acknowledgments

As editor, compiler and contributor to this anthology, my gratitude is as far reaching and varied as the chapter-essays presented here. Some of the gratitude takes on veiled dimensions that should remain so. Ostensibly, however, I thank teachers and colleagues who inspired or aided the project, perhaps in most cases unknowingly. These include Sy Ginsburg, Koschek Swaminathan, Don Conte, Albert Amao, Lillian Mayer, Mary Brandis, Mike Machiopa, Mem Masnick and others. I am also grateful to Leon Fair, Lew Ross, and Leonard Jackson for their "rose insights" and to Vernon O'Meally for overall administrative assistance to advance my chapters. One of my former graduate students, Lisa Cucciniello, contributed more than her own chapter on the rosary in the Catholic tradition; she also provided editorial and research assistance to a couple of other contributors. Likewise, Alexsandr Prodovikov was a faithful research assistant whose service was invaluable and steadfast over the span of the project. The Jacques Marchais Museum of Tibetan Art in Staten Island, New York, made available rare metaphysical books from the private collection of Jacques Marchais that were exceedingly helpful.

On a trip to Osaka in the summer of 2007, I had the good fortune of meeting Hisae Ogawa who introduced me to the work of Dr. Tomin Harada while we were touring Japanese cultural sights. The anthology is enriched, it is certain, by my overview of Dr. Harada's medical work to help the victims of the 1945 bombing of Hiroshima and the victims of agent orange exposure during the Viet Nam War. We applaud the unique rose garden movement begun by Dr. Harada in Japan before his death. Perhaps it is not inappropriate to also thank my late Dad, Frank Pauling, for he taught me much about the workings of the world in parable and metaphor as well as by his example of hard work, service and belief in family. During my pre-teen and teenage years he reminded me regularly that what I didn't know would "make whole new worlds." Of course, it was not at all clear what he meant by that until now. The lessons he taught become clearer and clearer with each passing year and every time I see a rose in full bloom. My parents' work to raise their children in the midst of rank racism, to serve their community as best they could and to stay married amid all sorts of negative outside forces, I can now see was a feat bordering on the phenomenal.

Finally, and most of all, abundant thanks to all who felt the mission of the rose and worked with me over the months, now years, that it took for the project to reach fruition. To all those who contributed essays or assisted directly in the anthology's production, including Albert Amao, Lisa Cucciniello and Alex Prodovikov in New Jersey, Mario Fenyo and Olga Mavlyutoza in Maryland, Monika Joshi in California, Tobe Levin in Germany, Godfrey Okorodus in Belgium, Leon Fair in Nebraska, Michael Price in Minnesota and Montgomery Taylor in New York City, I say thank you in the spirit of the rose.

Frankie Pauling Hutton, 2007 / 2012

Lightning Source UK Ltd.
Milton Keynes UK
UKOW05f0612180813

215506UK00001B/28/P

9 783981 386394

Special Limited 2012 Edition

ROSE LORE

ROSE LORE: ESSAYS IN CULTURAL HISTORY AND SEMIOTICS

By Frankie Hutton

UnCUT/VOICES Press

ISBN: 978-3-9813863-9-4
Bibliographic information published by the Deutsche Nationalbibliothek.
The Deutsche Nationalbibliothek lists this publication in the Deutsche
Nationalbibliografie; detailed bibliographic data are available on the
Internet at http://dnb.d-nb.de
Frankfurt am Main: UnCUT/VOICES Press, 2011.
Original: Frankie Hutton, ed. Rose Lore. Essays in Cultural History and
Semiotics. Lexington Books. A Division of Rowman & Littlefield Publis-
hers, Inc.: Lanham, MD, 2008.

UnCUT/VOICES PRESS

Martin Luther Str. 35, 60389 Frankfurt am Main, Germany

Tobe.levin@uncutvoices.com

www.uncutvoices.com
Geschäftsnummer HRB 86527, U.G. Haftungsbeschränkt